DURIAN
King of Tropical Fruit

Suranant Subhadrabandhu
and Saichol Ketsa

Daphne Brasell Associates Ltd, with Lincoln University Press
and
CABI *Publishing*

First published 2001 in New Zealand
and Australia by
Daphne Brasell Associates Ltd, with Lincoln University Press
PO Box 12 214, Thorndon, Wellington,
Aotearoa New Zealand
http://www.brasell.co.nz

and in the rest of the world by
CABI Publishing, CAB International,
Wallingford, Oxfordshire,
OX10 8DE, UK
and CABI North American Office,
875 Massachusetts Avenue, 7th Floor,
Cambridge, MA 02139, USA
http://www.cabi.org

A catalogue record for this book is available from the British Library, and from the Library of Congress, Washington DC, USA.

ISBN-13: 978-085199-496-3
ISBN-10: 0-85199-496-2

First published 2001
Transferred to print on demand 2010

Edited by Louise Armstrong, Paulette Crowley, Jo Fisher, Heather Gamble, Joan Halliwell, Jane Hingston, Frances Kinnaird, Amy Palmer, Alison Wilson and Sven Wilson, Whitireia Publishing, Auckland

Production supervision by Helen Badger, Rachel Clare, Vivianne Douglas, Eleanor Fearn, Nigel Heslop, Vivienne Hill, Caroline List, Daniel Platt, Paul Slatin, Stephen Towers, Rebecca Vuetilovoni and Tracey Wogan, Whitireia Publishing, Auckland

Designed by Hamish Thompson, Daphne Brasell Associates Ltd, Wellington

Cover photograph by Dr Suranant Subhadrabandhu

Indexed by Diane Lowther, Auckland

Typeset by Chris Judd, Auckland

Printed and bound in the UK by
CPI Antony Rowe,
Chippenham and Eastbourne

Contents

List of Figures and Plates

List of Tables

Foreword

We wish to recommend the book *Durian – King of Tropical Fruit* by Suranant Subhadrabandhu and Saichol Ketsa.

The book offers a comprehensive synthesis of work on durian, containing well-integrated coverage ranging from taxonomy to production and post-harvest management of the fruit. Specific areas covered include distribution, soils, fertilisers, tree management, pests and diseases, and harvesting.

We believe it will be of value to the small grower or harvester as well as the large producer, and will become the standard textbook for this fruit.

David Jackson
Professor
Department of Plant Science

Eric Scott
Group Leader
Ecology and Entomology Group

Soil, Plant and Ecological Sciences Division
Lincoln University
Canterbury
New Zealand

Preface

Durian is one of the most famous and controversial fruits in the world. It is highly valued and desired throughout South-East Asia. Although considered a delicacy by many, its fame is largely due to its unique, indescribable odour. The name 'king of tropical fruit' refers to two facets of the fruit: its superlative flesh, which is highly nutritional; and its appearance, which resembles the thorny thrones of the Asian kings of old.

Only a limited amount of research has been undertaken on the durian fruit. In *Durian: King of Tropical Fruit*, the authors have attempted to gather all the available information on this popular fruit. While foreign publications were surveyed and cited as additional resources, the authors acknowledge that the findings from these sources are not always applicable to the South-East Asian fruit-growing region. Much of the information contained in this book has been based on the experiences of the authors, and their research in Thailand and other countries of South-East Asia.

Detailed and comprehensive books on tropical fruits are rare and often hard to come by. The need for such works, suitable for students, horticulturalists and general readers, is clear.

This book is intended to be a resource for those who have an interest in durian and its cultivation, and those interested in the associated technology and research. It will also serve as an invaluable reference tool for government bodies who need up-to-date information on the economic potential of this fruit crop.

The authors are indebted to all fruit researchers, past and present, who have contributed much of the valuable information, without which this book could not have been written.

The authors also wish to express their sincere thanks to Mr Paphadipongse Vejachiwa of Canthaburi province and Mr Somjin Palakul of Rayong province – two outstanding durian growers in Thailand; Mr Sookwat Chandraparnik and Miss Surmsuk Salakpetch of Canthaburi Horticultural Research Center, Thailand; Dr Hiran Hiranpradit of the Department of Agriculture, Ministry of Agriculture and Co-operatives, Thailand; Dr Chan Ying Kwok of Fruit Research Division, MARDI, Malaysia; Dr T. K. Lim, formerly of the University of Pertanian, Malaysia; Dr Eric Scott, Lincoln University, New Zealand; and the student editors of Whitireia Publishing 1999 and 2000. Without their kind assistance this book would not have been possible.

Suranant Subhadrabandhu
Saichol Ketsa

This book is dedicated to all durian growers in Thailand and Malaysia

Part 1

The Plant and the Environment

1

History, Area of Origin, Nomenclature and Genetic Diversity

Introduction

The durian (*Durio* species), one of the most popular and famous seasonal fruits in South-East Asia, has been cultivated for centuries in tropical Asia. Durian probably first became known to the West during the sixteenth century, as recorded in the writings of the Dutch botanist, Linschoten, who wrote, 'The durian is of such excellent taste that [it] surpasses all other fruits in the world' (Barrett, 1912). Much later, the great naturalist A.R. Wallace wrote, 'To eat durian is a new sensation worth a voyage to the East to experience' (Ochse *et al.*, 1961; Wester, 1912). Indeed, the flavour of durian is so excellent that it is highly prized in all the areas in which it grows (Mortensen and Bullard, 1964). To the novice and many Westerners, however, its smell is nauseous and offensive (Brown, 1954). For this reason the durian is sometimes called the skunk of the orchard, and its fruit are barred from aeroplanes and most international hotels, even in the countries where it is produced (Anon, 1979). However, many people find that, despite its unusual flavour and odour, a taste for durian can be gradually acquired and may eventually become a passion.

Durian is a native of South-East Asia. Its centre of diversity appears to be the Malay Peninsula, Indonesia and the island of Borneo (Wilson, 1954; Reksodihardjo, 1962; Mortensen and Bullard, 1964; Anon, 1979).

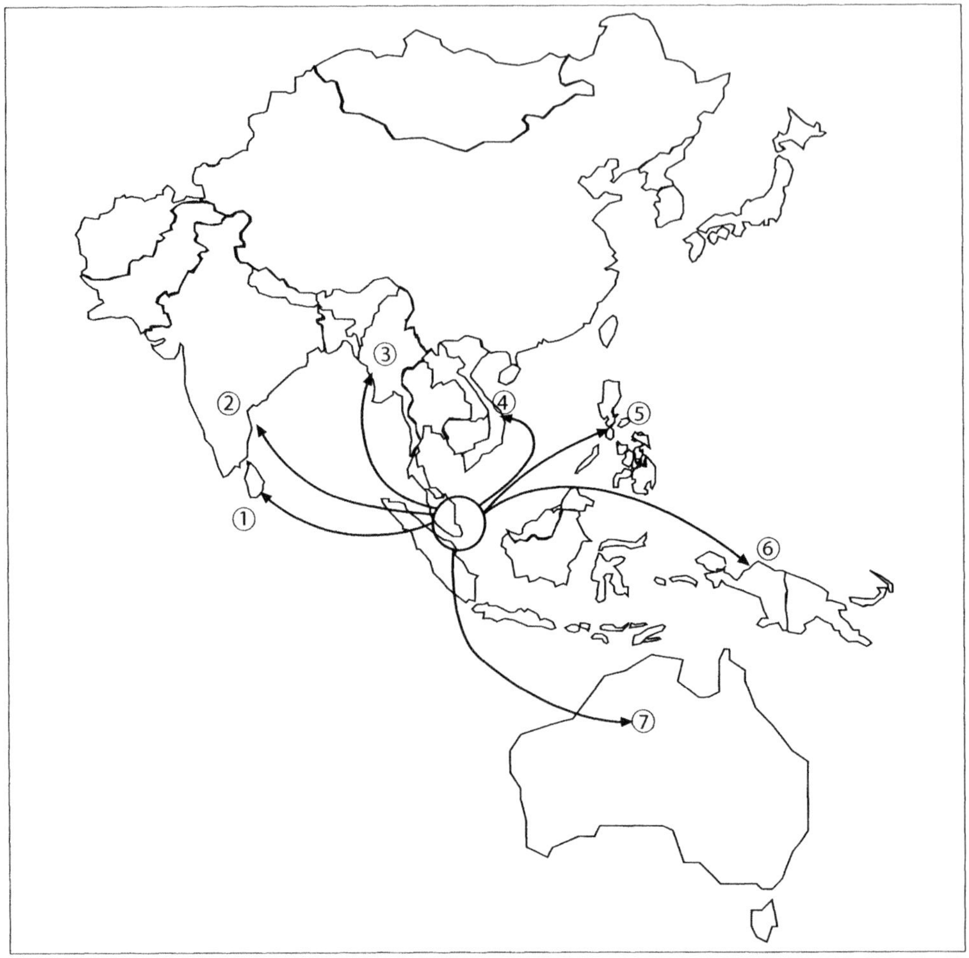

Figure 1.1 *Areas of concentration of cultivated types and sites suitable for in situ conservation of durian*

○ *Origin provenance: Southeast Asia, especially South Thailand, Malaysia, Borneo and Sumatra. Spread to:*

1. *Sri Lanka*
2. *South India*
3. *Myanmar*
4. *Vietnam*
5. *Philippines*
6. *New Guinea*
7. *Australia*

The most domesticated species of durian is *Durio zibethinus* Murr., which is distributed throughout the Malay Peninsula, southern Thailand, Malaysia, Indonesia, the island of Borneo and New Guinea, and has also been introduced to India, Sri Lanka, Myanmar, Vietnam and Jamaica (Figure 1.1). Several wild specimens have been collected in Sabah and there seems little doubt that it originated on the island of Borneo. *D. kutejensis* has its centre in East-Indonesian Borneo but is also cultivated near Kalabakan, Tawau in Kampong Pantai, Sipitang and on the Klius Peninsula of Indonesia. *D. oxleyanus*, easily recognised by its soft hairy leaves, is valued in and near villages as a fruit tree. *D. graveolens*, the red fleshed durian, has fruit which open on the branches of the tree and seeds surrounded by a red aril (flesh). This species is more common near villages than in the forest and is becoming more and more domesticated. *D. dulcis* is one of the more attractive durian, but has a strong odour that can be smelt from far away. The fruit is still unopened when ripe and the aril is yellow.

Nomenclature and Taxonomy

The durian is a tropical fruit tree in the order Malvales, family Bombacaceae, and is related to kapok. The family Bombacaceae is a strictly tropical family of plants native to both hemispheres. The most common and economically valuable species is *D. zibethinus* Murr. This species is also known by the synonym, *D. accuminatissima* Merr. Popular names for *D. zibethinus.* are: durian (English); in Indonesia, ambetan (Javanese), kadu (Sundanese), and duran (general); in the Philippines, dulian (Sulu); in Myanmar, du-yin; in Cambodia, thu-reen; in Laos, thourien; in Thailand, thurian (general) and rian (southern peninsula of Thailand, northern Malaysia); in Malaysia, duren; and in Vietnam, saurieng.

There are 51 genera and approximately 200 species within the family Bombacaceae (Reksodihardjo, 1962). The genus *Durio* has approximately 28 species, of which 19 are found on the island of Borneo and from this group of 19, 14 species are found in Sabah (Reksodihardjo, 1962) and 16 in Sarawak (Abang M. Mokhtar, 1991). In Malaysia there are 23 *Durio* species, of which 13 are found in the lowland forests of Peninsular Malaysia (Cockburn, 1976). Among the species found in Peninsular Malaysia, five are endemic. Seven species are known on the island of Sumatra but none of them are endemic (Kostermans, 1958). Four species are reported in the Philippines (Coronel, 1983), and six species are known in Thailand (Smitinand, 1980).

Related Genera

Genera in the family Bombacaceae that are botanically related to durian and produce edible fruits include: *Patinoa almirajo* Cuart., which has an elongated ovoid fruit and seeds that are completely covered by a fleshy, odourless, edible aril; *Pa. sphaerocarpa* Cuart. which has a spherical fruit and seeds that are also within an edible aril; and *Quararibea cordata* (H. & B.) G. B. & H., commonly known as 'sapote' in the Amazon Valley (Reksodihardjo, 1962).

The family Bombacaceae also includes some other economically important plants, such as those belonging to the genera: *Ceiba*, *Bombax*, and *Salmalia*, which produce kapok (silk cotton); *Ochroma*, which is the source of balsa, the lightest wood; *Adansonia*, which produces edible fruit and wood fibre for paper making; and *Pachira*, which produces a valuable seed oil (Stanton, 1966).

Related Species

Of the 28 *Durio* species, only seven besides *D. zibethinus* produce edible fruit (Zainal Abidin *et al.*, 1996). They are the common durian *(D. zibethinus)*, lai *(D. kutejensis)*, keratogan (*D. oxleyanus*), tabelak (*D. graveolens*), lahong (*D. dulcis*), kura-kura or turtle durian

Table 1.1: Distribution of Durio species in South-East Asia

Species	Pen. Mal	Sabah	Sara-wak	Kali-mantan	Suma-tra	Java	Myan mar	Philip pines	Thai-land	Viet-nam
D. acutifolius		+	+	+						
D. affinis		+		+						
D. beccarianus				+						
D. carinatus	+		+		+					
D. crassipes		*								
D. dulcis		+		+				+		
D. excelsus				*						
D. grandiflorus		+	+							
D. graveolens	+	+	+	+	+			+		
D. griffithii	+	+	+	+	+			+	+	
D. kinabaluensis		*								
D. kesudinarum				*						
D. kulejensis		+		+		+		+		
D. lanceolatus		+	+	+						
D. lissocarpus				+						
D. lowianus	+				+				+	
D. macrolepis	*								+	
D. macrophyllus	*									
D. malaccensis	+				+				+	
D. monsoni							+		+	
D. oblongus			+							
D. oxleyanus	+	+		+	+					
D. pinangianus	*									
D. purpureus		+		*						
D. singaporensis	*									
D. testudinarum		+	+	+						
D. wyatt-smithii	*									
D. zibethinus	+	+	+	+	+	+	+	+	+	+

**These species have been reported in horticultural literature but are no longer found in their natural habitats.*

(*D. testudinarum*), munjit or monyet (*D. grandiflorus*), and chaarian (*D. lowianus*) (Polprasid, 1983). The remaining 21 species are wild and inedible. The distribution of each species within South-East Asia is shown in Table 1.1. The 17 species found in Malaysia are shown in Table 1.2. At least six species are reported wild in Thailand (Table 1.3, Figure 1.2) (Smitinand, 1980).

The semi-cultivated edible species are *D. lowianus* (endemic to Peninsular Malaysia), *D. graveolens*, *D. oxleyanus*, *D. kutejensis* and *D. dulcis*. Wild species such as *D. kesudinarum* and *D. grandiflorus* also produce edible fruit but these are of poor quality.

Edible Species

In addition to *D. zibethinus*, seven other species in the *Durio* genera are reported to be edible. Descriptions of these species, including *D. zibethinus*, are as follows.

D. zibethinus Murr

This is the common durian that bears fruit of very high quality and economically is the most important species. Because it is heterogeneous, it exhibits a wide range of characteristics in

Table 1.2: Durio species in Malaysia

Species	Vernacular name	Distribution
D. affinis	–	*Johore, Trengganu, Sabah*
D. carinatus	*Durian paya*	*Johore, Pahang, Sarawak (restricted to peat swamp forest)*
D. griffithii	*Durian tupai/kaya*	*Sabah, Sarawak, and all states in Peninsular Malaysia, except Perlis and Penang*
D. lowianus	*Durian daun*	*Present in all states, except Perlis, Penang and Malacca*
D. macrophyllus	*Durian daun besar*	*Kelantan, Pahang, Perak*
D. malaccensis	*Durian batang*	*Trengganu, Selangor, Negeri Sembilan, Malacca, Johore*
D. pinangianus	*Durian daun tajam*	*Penang, Perak*
D. singaporensis	*Durian bujur*	*Trengganu, Pahang, Negeri Sembilan, Johore*
D. wyatt-smithii	*Durian hijau laut*	*Common in Bukit Bauk, Trengganu, Kelantan, Pahang*
D. zibethinus	*Durian*	*Throughout Malaysia*
D. testudinarum	*Durian kura-kura*	*Interior islands of Borneo, Sabah and Sarawak*
Minor Durio species		
D. acutifolius	*Durian maias*	*Sabah, Sarawak*
D. cassipes		*Sabah*
D. lanceolatus	*Durian burung*	*Sabah, Sarawak*
D. oblongus		*Sarawak*
D. purpureus		*Present in Sabah, but very rare*
D. kinabaluensis		*Present in Sabah, but very rare*

Table 1.3: Durio species in Thailand

Species	Vernacular name	Distribution
D. griffithii	*Thurian nok*	*Yala*
D. lowianus	*Chaurian, Thurian nok*	*Ranong, Trang*
D. malaccensis	*Due-yae buu-kong, Thurian don*	*Yala, Narathiwat*
D. monsoni	*Thurian thuean, Thurian paa*	*Chumphon, Phangnga*
D. pinangianus	*Thurian paa*	*Yala, Narathiwat*
D. zibethinus	*Thurian, Rian*	*General*

tree form, fruit form and aril quality amongst a population of seedling trees. Fruit colour ranges from green to crimson-yellow and the aril ranges in colour from white or cream to deep golden-yellow, and even orange-yellow. The seed can be small or large, and is chestnut brown. This highly variable species also exhibits differences in flowering and fruiting seasons, as well as fruit bearing ability.

D. testudinarum Becc

This is locally known as the turtle durian or kura-kura durian. The name 'turtle durian' may come from the fact that the tree bears fruit close to the ground where turtles can easily eat the ripe ones. This species is different to most *Durio* species in which fruit are generally borne on

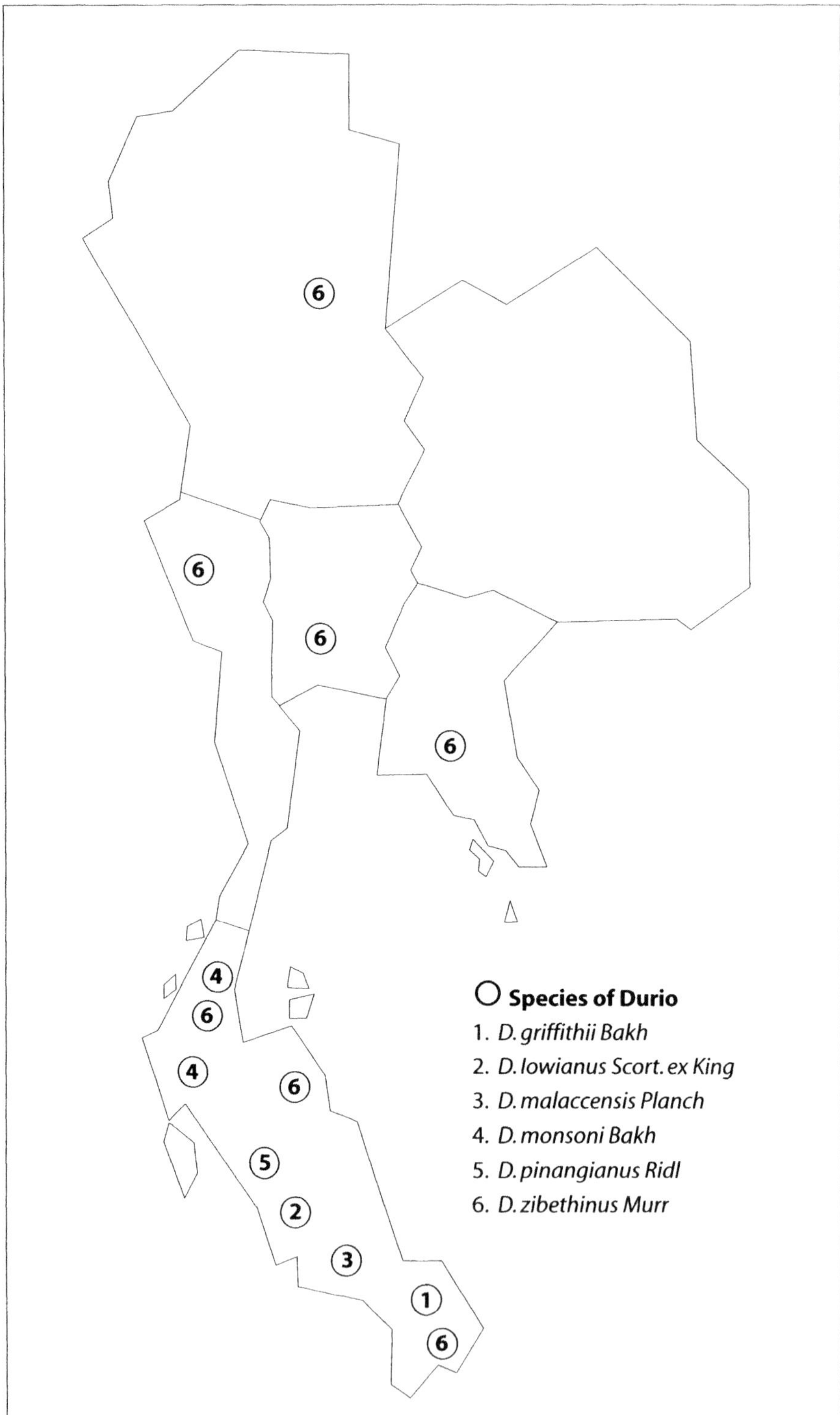

Figure 1.2 *Distribution of wild species of durian in Thailand*

primary and secondary branches. In this species, flowers and fruit are borne on the massive buttresses surrounding the trunk. This makes the heavy spined fruit easy to pick and less dangerous to pedestrians. This species exhibits very little variation, probably due to its self-pollinating nature. The trees have an extended flowering season. The fruit are ovoid with an impressive armour of sharp stout spines. They are greenish-brown when young and become yellowish-brown as they mature. The fruit contain five compact segments of light yellow aril, which surrounds large ovoid seeds. The aril is sweet, with a strong lingering aroma which can be nauseating. This species is rare and is found mainly in the interior valleys of Borneo amidst other large forest trees. It seems to prefer the damp humidity of the hot lowland conditions.

D. oxleyanus Griffith

This species is locally known as the beludu durian. It is another rainforest durian, found growing wild in Borneo, Sumatra and Peninsular Malaysia. It is an homogenous species that grows into huge trees in its natural habitat. The tree is easily recognised by its soft hairy leaves. *D. oxleyanus* is a valued fruit tree when it grows in or near villages, and is therefore cultivated in a small way by local village people. Whether cultivated or wild, the fruit are avidly eaten wherever they are found, and piles of discarded seeds occasionally create small groves of trees around settlements. Unlike most other *Durio* spp., which bear fruit with five locules, fruit of this species are characterised by only four locules. The small, round, grayish-green fruit are covered in stiff pyramidal spines. The aril is pale yellow, smooth and sweet with a fragrance similar to the common durian. There is limited clonal selection in this species.

D. kutejensis (Hassk.) Becc

This is locally known as the pulut durian. The species exhibits little variation in vegetative, floral and fruit characteristics, and bears fruit late in the season. The flowers are crimson with the corolla subtended on a stout receptacle. All other characteristics of the tree and fruit, even the taste of the aril, are similar to the common durian. The flowers produce a strong, lingering smell that attracts insects. The fruit are small and green with a dull yellow tinge. They are easily split and have soft spines. The aril is golden with a texture that is dry but smooth and creamy. Flavour and fragrance are milder than the common durian. The seeds are small and dark brown. Several clones of this species have been developed, of which at least one has a mild and inoffensive odour. The plant is native to the forests of Borneo. It can also be found along the foothills of the central ranges of Kalimantan, Sarawak, Sabah and Brunei Darussalam. A few trees have been planted in some neighbouring areas and exploratory plantings can be found in several parts of Indonesia outside of Borneo. It is sometimes cultivated in kampongs but even there the plants are more wild than domesticated. In Kalimantan, the pulut durian is already cultivated in the mixed dipterocarp forest. This could point the way to future rainforest replenishment activities.

D. dulcis Becc

This is a very rare species, locally known as the tutong durian. It is found growing wild only in the interior forests of Borneo. The tutong durian is occasionally cultivated in both Sabah and Kalimantan. It is regarded as the most attractive durian because the fruit is beautiful, with long, thin, red spines with a dark tip. The pale yellow aril is both fragrant and sweet, probably the sweetest of all durian fruit. The texture is soft and creamy like a good caramel, and it is said to have the flavour of pineapple without the acidity. The fruit is difficult to open and is usually chopped in half with a machete. The rind can have a strong smell, but the flesh is far less overpowering than that of the common durian. The seeds are shiny and black. Both the seeds and the aril are said to have aphrodisiac properties.

D. graveolens Becc

This species is locally known as king durian, or durian kuning. The fruit are almost odourless and come in a range of bright colours. The colour of the aril is extremely attractive – it is normally dark red (described as 'bougainvillea red'), but can also be orange ('mango orange'), brown ('baked bread brown'), or yellow ('maize'). The rind of the fruit is orange-yellow with sharp pyramidal spines. The rind has very little odour. This species is indigenous to Borneo, and is found growing wild in the jungles of eastern Borneo, Sumatra and Peninsular Malaysia (at altitudes up to 1 300 m). It is occasionally found cultivated. The king durian is, like the common durian, heterogeneous. The red aril is sweet and cheesy, with a fragrance of roasted almond in fruit from certain trees, whilst in others the taste is milder and less lingering. Several good clones of this species have been developed in Sabah. The distinctive aril is characterised by its very thick creamy texture and lack of sweetness (similar to avocado) that gives it the potential for use in savoury dishes. It can be eaten raw, or cooked with onions and seasonings and served with rice and vegetables. The king durian differs from other *Durio* spp. in that, as it does not drop its fruit when they are ripe, they must be picked from the tree.

D. grandiflorus (Mast) Kosterm or munjit durian

This species is found in Sabah, Sarawak and Kalimantan, where its natural habitat is in rainforests at altitudes up to 500 m. The trees are occasionally cultivated because their fruit are edible. The fruit have stiff spines up to 2 cm long and the aril is yellow.

D. lowianus Scort. ex King

This species is found in the swamp forests of southern Thailand, Peninsular Malaysia and on the island of Sumatra. It is known in Thailand as 'Chaarian' (in Ranong province) and 'Thurian nok' (in Trang province). The trees are occasionally cultivated because of their edible fruit. However, the round fruit only have a thin layer of flesh which is of inferior eating quality.

Phylogenetic Relationships

The genetic centre of the genus *Durio* is in Borneo, where at least 19 species are distributed and some minor species are cultivated on a small scale. Although *D. wyatt-smithii* Kosterm is thought to be the wild ancestor of the cultivated durian (Sudhadrabandhu *et al.*, 1991), the phylogenetic relationships among the species producing edible fruit are not clearly defined. Kanzaki *et al.* (1998) studied the phylogenetic relationships between the common durian (*D. zibethinus* Murr.) and the six other edible species, grown under the same conditions at Plew Horticultural Research Station, Chanthaburi province, Thailand. Using the RFLP analysis of an amplified region of cpDNA, they found that *D. zibethinus* formed a monophyletic group with *D. griffithii* in the unordered parsimony tree. No restriction site mutation was detected between these two species and only two length mutations appeared in the amplified region of cpDNA, indicating a very close relationship. *D. griffithii* has potential value as a rootstock for *D. zibethinus* due to its resistance to disease. Although *D. griffithii* has not been given much attention until now, it may have contributed to the evolution of *D. zibethinus* (Kanzaki *et al.*, 1998). Unfortunately, in the study of Kanzaki *et al.* (1998), researchers could not include *D. wyatt-smithii*, which is thought to be the wild ancestor of the cultivated durian (Subhadrabandhu *et al.*, 1991). Therefore, an immediate common ancestor of *D. zibethinus* could not be defined.

In the same study of edible fruit-producing *Durio* spp. Kanzaki *et al.* (1998) found two groups in the phylogram (apart from *D. zibethinus*). *D. graveolens*, *D. kutejensis* and *D. monsoni* formed one group, whereas *D. lowianus* and *D. oxleyanus* formed another group (Figure 1.3). Although all these species are cultivated for local markets as fresh fruit, no close relationships between these species to *D. zibethinus* were found. Therefore, Kanzaki *et al.* (1998) suggested that these edible *Durio* spp. had evolved independently of *D. zibethinus*.

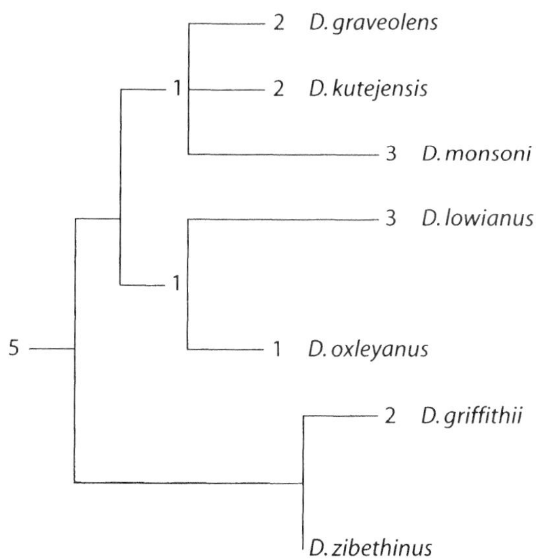

Figure 1.3 *The most parsimonious tree from unordered RFLP analysis (branch and bound search). The numbers represent branch length (Kanzaki et al., 1998)*

Genetic Erosion

There is an increasing concern in many tropical countries, including South-East Asian countries, about the loss of genetic diversity in plant species, especially those found only in this part of the world. Durian, which is found native only in South-East Asia, is also in danger of genetic erosion. Rapid deforestation clearing the way for economic development constitutes the most serious single factor contributing to genetic erosion; deforestation may lead to the destruction of the natural habitat of specific *Durio* spp. and hence their variability could be seriously affected. With the exception of *D. zibethinus*, the genetic diversity of the *Durio* species is difficult to quantify.

Malaysia, and to a lesser extent Indonesia, still have large tracts of natural tropical forests which are the native habitat of *Durio* spp. In Malaysia various forest reserves have been created to protect valuable forest species, including the durian. In contrast, some of the *Durio* spp. have already disappeared in countries such as Thailand and the Philippines. Aside from deforestation, the widespread cultivation of a few selected commercial clones is replacing the traditional varieties of durian in most Asian countries. In Malaysia and Indonesia, traditional varieties and farmers' selections still constitute a significant component of durian production, whereas in Thailand the local durian industry is mainly limited to a few recommended varieties. Certain *Durio* spp., such as *D. carinatus* or 'durian paya', thrive only in a swampy environment and are endemic to Peninsular Malaysia. If natural swamp forests are destroyed, then their survival is threatened (Zainal Abidin *et al.*, 1996).

2

Areas of Production and Economic Importance

Being a tree of the humid tropics, the durian needs high relative humidity throughout the year and as a forest tree, it grows better in shade and under shelter when young. When fully grown, it becomes a very large tree. In the early days, durian trees were grown near the riverside settlements in the Malaysian regions of Perak, Johore and Pahang, several provinces around Bangkok and parts of southern Thailand. Durian grows in most parts of Indonesia and in the Davao province of Mindanao in the Philippines.

Past attempts to introduce durian to other tropical areas of the world have generally failed. This is partly due to the fact that the durian seed quickly loses its viability during shipment. Durian cultivation on a small scale is reported in Africa, on the island of Madagascar in particular, where the crop was introduced by Malay immigrants. From Madagascar, its cultivation spread into the Congo of mainland Africa. In recent years, there has been some effort at commercial durian cultivation in northern Queensland and the Northern Territory of Australia. The importation of durian into Australia began in 1975 and orchard planting started in 1980. Some durian have been planted in Florida but this is still limited to experimental testing. Some interest in cultivating durian has started in Hawaii.

Malaysia

Durian is grown in most of the Malaysian states. The main areas for durian cultivation include south-west Penang; Parit; Kuala Kangsar and Batang Padang in Perak; Jasin in Malacca; and Segamat, Muar and Batu Pahat in Johor. The total durian production area in Peninsular Malaysia was 21 558 hectares in 1980 and 31 225 hectares in 1985, after which it increased

Figure 2.1: *Major durian cultivation areas in Peninsular Malaysia (MARDI, the Malaysian Agricultural Research and Development Institute)*

sharply to 62 153 hectares in 1991. Relative to total fruit cultivation, the durian area was 18.4% in 1980, 25.4% in 1985 and 31.0% in 1992 (Table 2.1). Durian is the most popular fruit in these regions and constitutes the largest percentage of the total fruit-growing area.

The majority of durian orchards are concentrated in Johore (36.4%), Perak (23.9%) and Pahang (11.1%) (Chan *et al.*, 1981) (Figure 2.1). The percentage of durian orchards in all other states is less than 10% of the total, and is as low as 1.1% in Perlis and the Federal Territory. Over 87% of the growers are indigenous Malays who own most of the old 'dusuns' (mixed home gardens); 11.7% of the growers are Chinese and 0.7% are Indians. This distribution appears to be similar over most states in Peninsular Malaysia except for Penang where 77.2% of the growers are Chinese. In Johore, the three prominent districts are Muar, Batu Pahat and Pontian while in Perak, the major fruit-growing districts are Larut Matang, Parit/Kuala Kangsar and Hilir Perak. Together, these six districts in Johore and Perak account for about 53% of the total durian-producing area in Malaysia (Figure 2.2).

Figure 2.2: *Ecological zones suitable for durian cultivation in Peninsular Malaysia (MARDI)*

***Table 2.1:* Durian hectarage in Peninsular Malaysia (1980–1992) (Ministry of Agriculture, Malaysia, 1993)**

Year	Hectares	Percentage of total fruit (hectarage)
1980	21 558	18.4
1981	20 604	18.2
1982	24 664	20.7
1983	25 771	22.4
1984	29 641	24.4
1985	31 225	25.4
1986	36 143	29.7
1987	42 070	29.4
1988	48 049	32.7
1989	54 665	33.7
1990	56 889	33.9
1991	62 153	32.0
1992	61 294	31.0

In Malaysia, durian production is increasing. In 1980, it was estimated that 134 600 tonnes of durian were produced, valued at US$73.9 million (RM201.9 million). Production rose to 389 900 tonnes in 1991, worth US$214.2 million (RM584.8 million) (Table 2.2). Most of the durian is produced for fresh consumption and consumed locally. Exports in 1992 were 26 890 tonnes valued at US$17.9 million (RM44.6 million). In 1989, 13.7 tonnes of durian were used as raw material for processing. Per capita consumption of durian in 1989 was estimated to be 4.16 kg/person/year (FAMA, 1989).

Table 2.2: Durian production in Peninsular Malaysia (Ministry of Agriculture, Malaysia, 1993)

Year	Volume ($x\ 10^3$ tonnes)	Value (million) US$	RM
1980	134.6	90.9	201.9
1981	128.3	86.7	192.4
1982	153.9	99.5	230.8
1983	160.1	102.6	240.1
1984	194.4	113.8	276.6
1985	224.9	139.4	337.3
1986	*	*	*
1987	262.1	156.0	393.1
1988	289.5	172.3	434.2
1989	319.7	177.6	479.5
1990	353.1	196.2	529.6
1991	389.9	214.2	584.8
1992	384.5	211.2	576.0

** data not available*

Durian cultivation in Malaysia is regarded as a small-scale operation characterized by smallholdings, 'dusun', of half to one hectare each, usually intercropped with other fruits. Until recently, Malaysia's agricultural development programme favoured the production of commercial crops such as rubber, oil palm and cocoa, over fruit. However, the new National Agricultural Policy that the government introduced in 1984 emphasised the fruit industry as a potential source of profit for growers, including small farmers from the private sector (Jamil Ahmad, personal communication).

Durian trees bear fruit once a year in most parts of Malaysia. However, environmental factors can affect the fruiting season, eg, two fruiting seasons occur in areas where there are two obvious dry seasons, such as on the west coast of Peninsular Malaysia. On the east coast, fruiting occurs only once a year.

The main fruiting season on the west coast of Peninsular Malaysia falls in November and December while the off-season is in May and June. (Normally less fruit is produced during the off-season.) On the east coast, the fruiting season is normally during August and September. These fruiting seasons can vary, or may not occur at all in a certain area, depending on climatic factors. In the southern and central states of the Peninsula, the season extends for three months with its peak in July. However, in the northern states, the season begins in July where it appears to be shorter – extending for two months with a later peak in August. In Sarawak, there are two seasons, one from September to October and the other from March to April.

Thailand

Durian has been grown in Thailand for over 300 years. It is believed that durian was introduced to central Thailand in 1787 from the south-east region of Myanmar (Burma) and to southern Thailand from Malaysia (Bhusiri, 1970). Many wild durian trees that are over 100 years old can still be found in southern Thailand (Punsri, 1972). Durian was recorded growing as a backyard crop in 1787 by some growers in the Thonburi district, near Bangkok. Later, small durian orchards were established by individual growers in the Nonthaburi province, west of Bangkok.

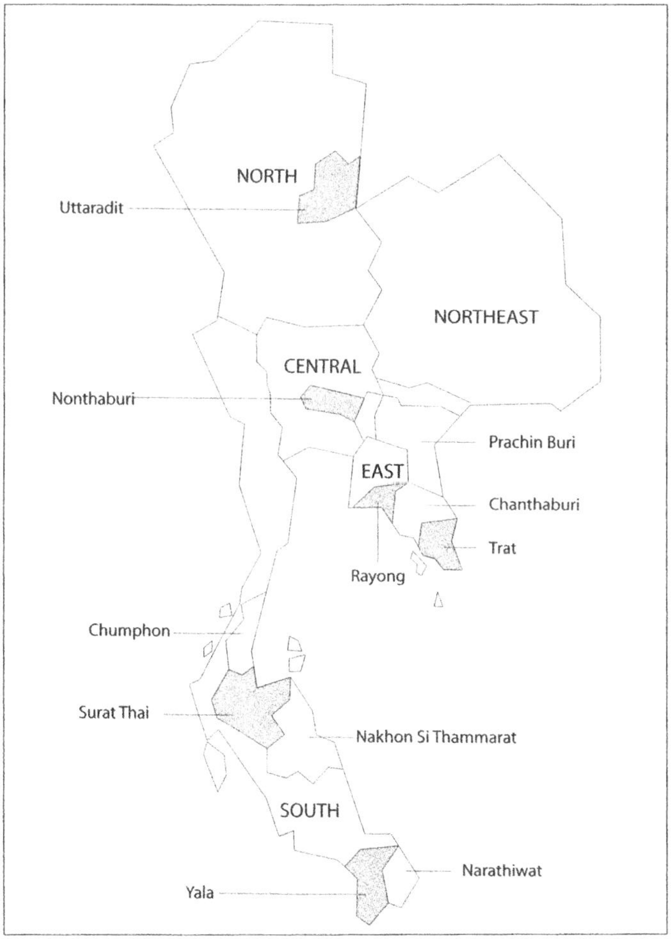

Figure 2.3: *Major areas of durian cultivation in Thailand (Nanthachai, personal communication)*

Many good durian cultivars grown near Bangkok were lost during heavy flooding in 1875 and 1942. Additional destruction by a hurricane in 1954, then a drought and subsequent salt water contamination of the soil in 1958, almost destroyed the durian industry in the Thonburi and Nonthaburi areas (Polachart, 1952; Chartjareon, 1961). In 1954, however, durian from Thonburi were successfully planted in the Chantaburi province in eastern Thailand. In following years, more durian trees were planted in Chantaburi and it is now the leading province for durian production. Other nearby provinces such as Rayong, Prachin Buri, and Nakhon Nayok have also grown durian cultivars from Thonburi and Nonthaburi (Bhusiri, 1970).

In 1962 and 1963, durian cultivars from Thonburi and Nonthaburi, such as Kan Yao, Mon Thong, Chanee, Kob and Chau-ma-fai, were used to replace the durian trees destroyed by a hurricane in southern Thailand in 1952. Of these, the cultivar which adapted most successfully to the environmental conditions in the south was Mon Thong (Rattanapruk, 1969).

Commercial scale durian production in Thailand is concentrated in the central region, parts of the north, and in the east and the south (Figure 2.3). The main growing areas are in the east and south, where the provinces with the highest production are Chantaburi (52.5%), Rayong (17.2%) and Chumphon (4.8%) (Subhadrabandhu, 1993). The other areas of production, which account for only 25.5% of total production, are Trat and Prachin Buri in the east, Surat Thai, Nakhon Si Thammarat, Narathiwat and Yala in the south and Uttaradit in the north (Table 2.3).

Table 2.3: The main durian-growing provinces, their planted area and production figures in Thailand in 1993 (Department of Agricultural Extension, Ministry of Agriculture and Cooperatives, Thailand, 1996)

Provinces	*Planted area (ha)*	*Volume produced (t)*
Chanthaburi	*37 136*	*392 312*
Rayong	*15 494*	*128 252*
Chumphon	*16 645*	*35 604*
Narathiwat	*4 011*	*25 343*
Trat	*5 172*	*32 757*
Uttaradit	*4 081*	*9 722*
Others	*29 521*	*122 652*
All provinces in Thailand	*112 060*	*746 642*

Between 1987 and 1994, the planted area of durian in Thailand increased from 76 140 to 115 593 hectares and durian production increased from 426 254 to 780 393 tonnes (Table 2.4). This is an increase of 51.8% in planted area and 90.1% in production volume.

Table 2.4:* *Area and volume of durian production in Thailand from 1987 to 1994 (Department of Agricultural Extension, Thailand, 1987–1994)

Year	*Planted area (ha)*	*Bearing area (ha)*	*Non-bearing area (ha)*	*Volume (t)*
1987	76 140	55 095	21 045	426 245
1988	84 500	58 311	26 189	444 145
1989	82 927	57 527	25 400	468 645
1990	90 340	55 316	35 024	464 959
1991	95 367	64 146	31 221	539 190
1992	97 238	64 633	32 605	711 371
1993	112 060	70 893	41 167	746 642
1994	115 593	74 712	40 881	780 393

Cultivars grown commercially are: Mon Thong (40.6% of the total durian plantation area), Chanee (32.9%), Kan Yao (6.0%), Kradum (2.4%) and others (18.1%). However, in terms of tonnage, Chanee is the leading cultivar, producing 36% of the total durian yield, followed by Mon Thong (31.8%) (Table 2.5).

Table 2.5:* *Area and volume of production of commercial durian cultivars in Thailand in 1991 (Department of Agricultural Extension, Thailand, 1992)

Cultivar	*Area planted (ha)*	*Area planted (%)*	*Bearing area (ha)*	*Bearing area (%)*	*Volume produced (t)*
Mon Thong	38 686	40.6	19 660	31.8	171 590
Chanee	31 412	32.9	23 816	36.0	194 092
Kan Yao	5 748	6.0	4 066	5.9	32 042
Kradum	2 230	2.4	1 776	2.4	13 044
Others	16 491	18.1	14 827	23.9	128 421
Total	94 491	100.0	64 145	100.0	539 190

Durian trees generally do not produce fruit for the first four to five years and take 15 to 20 years to reach maximum production. A large area was planted with durian in Thailand in 1994 and therefore the production of durian in Thailand is expected to increase substantially.

Harvesting in Thailand begins in late March and continues until September, with peak production from May to July. This extended harvesting season is due to the diversity of varieties grown and growing locations. Fruit from the eastern provinces are harvested and marketed from April to June; from July to September in the south, and from August to early October in the north. Early season fruit generally fetch high prices. The price in the off-season can be as high as US$8.00 per kg.

Indonesia

The major durian production areas in Indonesia are shown in Table 2.6. The area planted, yield and volume of production of durian from 1987 to 1993 are shown in Table 2.7.

Table 2.6: Major durian production regions in Indonesia in 1992 showing area planted, yield and volume of production of durian by region (Directorate General of Food, Crops and Horticulture, Indonesia, 1994)

Region	Area planted (ha)	Yield (kg/ha)	Production (t)
Sumatra	14 062	5 193	73 098
Java	12 865	4 663	59 992
Kalimantan	4 663	2 175	10 144
Sulawesi	2 538	879	2 231
Bali	875	6 479	5 669
Maluku	556	1 254	697
Nusa Tenggara	451	820	603
West Irian	14	4 786	67

Table 2.7: Area planted, yield and volume of production of durian in Indonesia, 1987–1993 (Directorate General of Food, Crops and Horticulture, Indonesia, 1994)

Year	Area planted (ha)	Yield (kg/ha)	Production (t)
1987	41 620	4 790	199 360
1988	36 054	5 359	193 200
1989	24 859	5 599	139 193
1990	43 756	5 544	242 585
1991	42 758	4 808	205 389
1992	36 024	4 233	152 501
1993	44 016	5 195	228 668

Until the late 1980s, most durian production in Indonesia came from mainly undetermined cultivars in backyards or in the forest near villages. In the early 1990s, thousands of hectares of new orchards were planted. These orchards, approximately 50 hectares per unit in size, are planted with only one cultivar, suitable for that area. In a few years these new areas will start producing fruit on a commercial scale

At present, most durian in Indonesia are grown in home gardens. These gardens are mainly in Java and Sumatra but durian trees are scattered all over the country, especially in Medan. The tree is usually grown under semi-jungle conditions as in the old mixed orchards. Under these conditions, durian trees grow very large, and the lower branches self-prune leaving bearing branches that are very high above the ground. The flowering and fruiting of these trees is erratic but because of the height of the branches, thinning of flowers and fruit is not practical (although it may be desirable). In general, the fruit produced from these home gardens are of low quality and are usually sold only at fruit stalls along the roadside during

January and February, with smaller amounts available for a further two months. It is likely that future durian orchards established in Indonesia will use more advanced technology, such as that in Thailand.

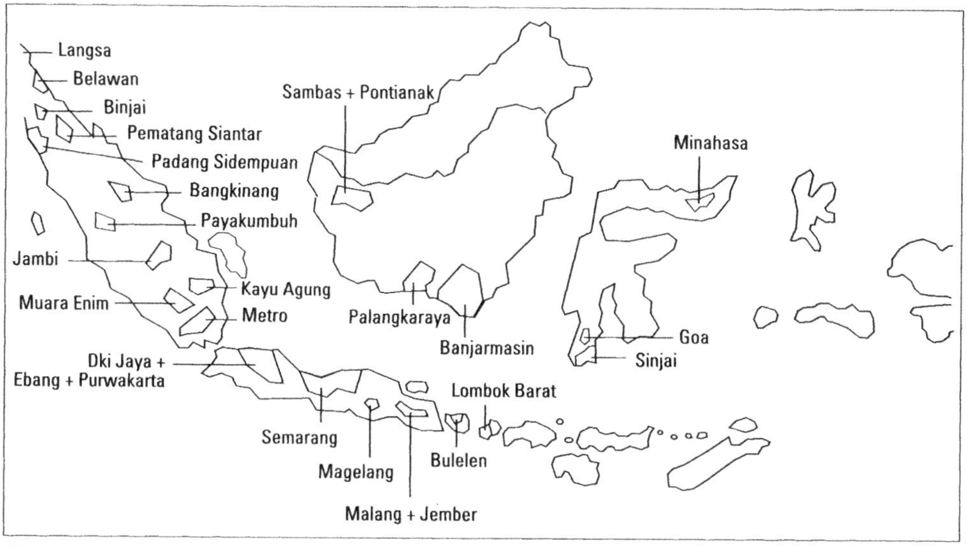

Figure 2.4: *Durian-producing areas of Indonesia*

Philippines

In the Philippines, durian is considered a minor fruit crop and ranks 16th in the planted area of locally grown fruits (Bureau of Agricultural Economics, 1977). In 1977, the area planted in durian was estimated to be 1 080 hectares; this consisted of 31 500 bearing trees and total production was 5 935 tonnes (Coronel, 1983). Production was mainly concentrated in the four regions of Mindanao, with southern Mindanao having the largest area planted in durian (720 hectares). However, the best tasting durian fruit grown in the Philippines comes from Davao City (and in particular, the Calinan District), 400 to 850 m above sea level with fertile loamy soil. The average annual rainfall in this area is 3 800 mm and the evenings are cool.

There are no updated statistics for durian production in the Philippines, although the presence of numerous durian nurseries indicates a significant increase in demand for asexually propagated plants of local clonal selections. This suggests that more people have started to grow durian as a monocrop, not only on a small scale, but also in commercial scale plantings. Flowering starts in April and continues until June; peak harvest is from August through to November (Coronel, 1983). An estimated area of 8 000 hectares was planted with durian in 1994.

Durian production in the Philippines is mainly for domestic consumption but not enough fruit are produced to satisfy local demand. Hence durian is one of the most expensive fruits there. Until 1993, the average wholesale price varied from US$1.06 – $7.71 per fruit. During the 1993 season, durian retailed from US$2.85 – $4.30 per fruit in Davao and US$6 – $20 in Manila. The variation in price reflects the seasonal availability of durian.

In the Philippines, ripe durian fruit are frequently eaten for dessert. The fruit is also processed into candy, jams and preserves, as well as fillings for rolls, tarts and pies. More recently, commercial processing of durian into flavouring for icecream and milkshakes has developed.

The people of Mindanao appreciate the strong pungent aroma and distinctive flavour of the local cultivars, but most other people, especially those not accustomed to the durian taste, prefer the milder aroma and flavour of Thai cultivars such as Chanee and Mon Thong.

Brunei Darussalam

Brunei Darussalam is known for its great diversity of tropical fruits – among these is the durian. Many species of durian are found on the island of Borneo but many other species are located in the regions around Brunei Darussalam Bay and the riverside settlements and interior enclaves of Brunei Darussalam. With the exception of a few commercial orchards, most durian is produced from the multitude of smallholdings. Fruits are gathered from 'pulau buah' (isles of fruit) scattered all over the country. These 'pulau buah' are concentrated areas of mixed fruits that were established by early settlers but are now abandoned and have reverted to a semi-wild state. During the fruiting seasons, farmers visit and camp in their 'pulau buah' where they clear the undergrowth and wait for the durian fruit to fall. This annual practice helps to maintain these areas, providing the durian trees with a competitive edge because the trimming of the undergrowth gives the durian trees room to grow and produce fruit.

The total fruit-growing area of Brunei Darussalam is approximately 1 000 hectares, with durian occupying 10% to 15%. This area includes backyards, home gardens, mixed orchards and 'pulau buah'.

In the smallholdings and 'pulau buah', indigenous species predominate. *Durio graveolens* has adapted to a wide variety of habitats while *D. kutejensis* is restricted to the interior lowlands. *D. oxleyanus* and *D. testudinarium* find their niche in the alluvial valleys and *D. dulcis*, the edible 'wild' durian, is widespread (Figure 2.4). However, it is uncommon to find *Durio* spp. in swampy marshes or peat swamps. A representation of the distribution of *Durio* spp. in Brunei Darussalam is shown in Figure 2.5.

Most of the indigenous species are hybrid trees that are decades old. These trees are a storehouse of genetic diversity for *Durio* spp.

In an attempt to upgrade the quality of durian in Brunei Darussalam, the Department of Agriculture began a monitoring scheme in 1985 to collect and register good quality durian, both of indigenous and introduced varieties (Department of Agriculture, 1988). Nearly 73 durian with good quality arils have been registered. Indigenous species make up much of the collection and the rest are introduced *D. zibethinus* cultivars. From this list, the Department of Agriculture has selected several indigenous varieties to be propagated vegetatively as Brunei Darussalam durian clones.

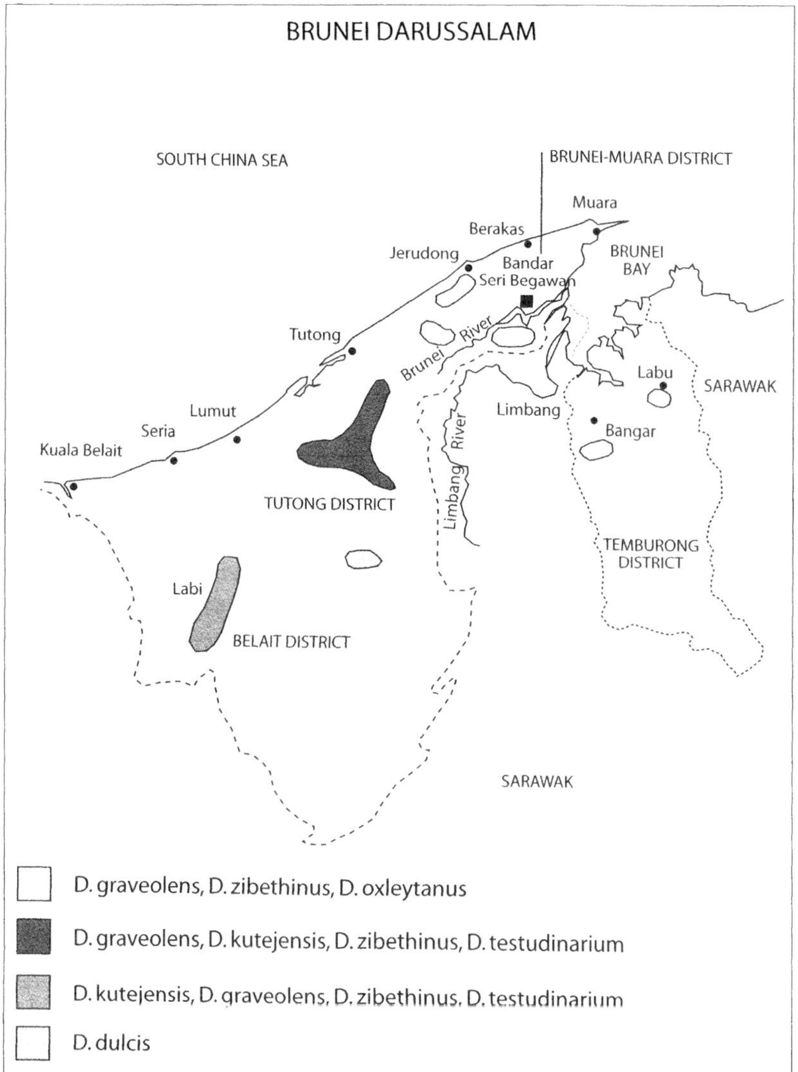

Figure 2.5: *Approximate distribution of Durio spp. in Brunei Darussalam (Department of Agriculture, Brunei Darussalam)*

The main fruiting season in Brunei Darussalam stretches from July to September with an occasional crop in January/February. Durian grown in Brunei Darussalam has enormous potential as a commercial fruit for niche markets. Tourism exposes people to a wide variety of fruits and offers the opportunity to acquire new tastes. However, there are limitations on exporting. For example, varieties such as *D. testudinarium* may never be sold in large amounts to Thailand where lovers of durian prefer *D. zibethinus*.

Singapore

Durian are grown in Singapore's agrotechnology parks; these parks are 1 700 hectares in size and have 12 hectares set aside for cultivating fruit trees. The major durian cultivars grown are 'H. C. Tan No. 2', 'H. C. Lim' and 'Lim Keng Meng'. These cultivars fruit from May to July and from November to January. Once trees of these cultivars reach 15 years of age, the annual yields are 26, 57 and 30 kg, respectively. The value of durian produced in Singapore is insignificant compared with the value imported. In 1992, durian imported into Singapore from Malaysia and Thailand was worth US$30 million, making it the most valuable imported fruit.

Economic Importance

The role of fresh and processed durian flesh in the economy of the ASEAN countries is expanding. This expansion is of tremendous benefit to growers, traders and everyone else involved in the trade. Among the ASEAN countries, Thailand is the largest producer of durian with 444 500 tonnes produced in 1987 and 1988 from a production area of 84 700 hectares. Indonesia recorded the production of 200 000 tonnes in 1985/1986, and in Peninsular Malaysia, 262 000 tonnes were produced in 1987 from a production area of 42 000 hectares. The Philippines produced only 16 700 tonnes from 2 030 hectares in 1987.

Between 1988 and 1994, Thailand increased its exports of durian by an average of 30% per annum and the value of these exports increased by 26% per annum. In 1988, Thailand exported 6 114 tonnes of fresh durian fruit valued at US$4.75 million. By 1992 this had risen to 15 116 tonnes worth US$11 million. Exports of processed durian also increased by an average of 56% per annum and the value of these exports increased by 27% per annum. In 1988, Thailand exported 635 tonnes of processed durian valued at US$3 million and by 1992, the quantity had increased to 2 399 tonnes valued at US$7 million. Among the countries that import fresh durian fruit from Thailand are Hong Kong, which imports 65% of Thailand's exported durian fruit, and the United States and France. Medium-sized fruit, weighing 1 to 2 kg each, are preferred for export because this is the size of durian with which foreign markets are familiar. Larger fruit are mainly sold locally and may fetch as much as US$4 to $5 each in the Bangkok and Chiang Mai markets if they are of superior cultivars.

The expansion of durian production has other benefits. Not only are labour requirements for planting and harvesting increased but there is also demand for associated products, such as agricultural machinery, harvesting tools, cartons, plastic trays, plastic buckets, cold storage facilities, fertilisers, pesticides and fungicides.

Nutritional Values

The aril represents 20% to 35% of the total fruit weight and the seeds 5% to 15%. The aril and seeds are both very nutritious, being rich in carbohydrates, protein, fat and minerals. The durian fruit consists of: water 66.8 g per 100 g edible portion; protein 2.5 g; fat 1.6 g; carbohydrates 28.3 g; fibre 1.4 g; calcium 20 mg; phosphorus 63 g; potassium 601 mg; iron 0.9 mg; sodium 1 g; thiamine (vitamin B1) 0.27 mg; riboflavin (vitamin B2) 0.29 mg; vitamin C 57 mg; and beta-carotene 10 mg. The energy value is 520 kJ per 100 g, which is high, compared with other fruits. The characteristic durian odour stems largely from thiols or thioethers, esters and sulphides.

Uses

Generally, the durian fruit is composed of 55% to 66% percent rind, 12% to 15% seeds and 22% to 30% aril (the edible part) (Ng and Thamboo, 1967). Usually, the aril is eaten fresh. In Indonesia, local villagers are so fond of the fruit that the rice harvest suffers if it coincides with the durian harvest and during peak seasons people prefer to eat durian when they are hungry. Today, fresh arils can be shrink-wrapped and deep-frozen to extend the supply season; in this form the fruit is gaining acceptance in the export trade.

Besides being consumed as fresh fruit, the aril can be processed in a variety of ways. It can be made into jam, candy, or used as icecream flavouring (Reksodihardjo, 1962). In Thailand, during the peak season, the aril is commonly preserved as durian cake. This is made by boiling the aril with flour and sugar and leaving it until it is partly dried. In Indonesia and Malaysia, the aril may be allowed to ferment and then eaten as a side dish, mixed with rice and sugar, or minced with salt, onion and vinegar (Reksodihardjo, 1962). The unripe or partly ripe aril may be used for making soup. Unripe fruit may also be boiled and eaten as a vegetable (Coronel, 1983).

The seeds are edible and are almost pure starch. Boiled or roasted, they are eaten as a snack (Monsalud *et al.*, 1966). In Indonesia, the seeds are roasted, cut into slices and fried in spiced coconut oil to be eaten with rice, or coated in sugar and consumed as a sweet (Reksodihardjo, 1962). Young shoots may be boiled and eaten as a vegetable while the durian flower petals make a good salad or can also be cooked as a vegetable dish.

When durian are plentiful, the arils are processed and preserved as various traditional products such as 'durian kuan' (the aril cooked with flour and sugar), 'tempok' (the aril cooked with sugar) or 'tempoyak' (salted aril) or modern products such as jams, toffees, tarts, icecream and milkshake flavouring.

The tree trunk, which can grow to a height of 30 m, is used for many purposes. The timber is white and tinted red or brown towards the centre, coarse and not durable. However, it is used for indoor construction and cheaper types of furniture. The rind of the fruit can also be dried and used as a fuel. In addition, the ashes of the rind can be used for bleaching silk.

Ethnobotanical Aspects and Local Knowledge

Due to the durian's long history of cultivation, various ethnobotanical beliefs about durian have emerged among the indigenous people of ASEAN countries, particularly in Malaysia and Indonesia. Such beliefs are seldom documented and no attempt has been made to verify them scientifically. Some examples of the local beliefs about durian are listed below:

- The durian flesh is an aphrodisiac; the fruit are much sought by those who believe this.
- A compound prepared from durian leaves and roots is therapeutic as a treatment for fever. The leaves are also therapeutic when used in a medicinal bath for jaundice, and when applied to epidermal swelling (Reksodihardjo, 1962).
- The rind of the fruit is often boiled to apply externally to skin ailments (Coronel, 1983).
- The ashes of the fruit-walls can be used to cure fever in infants.
- The durian tree should be planted during the full moon if it is to bear heavily in later years.
- One should not point one's fingers at the durian inflorescence or else the fruit will not set.
- 'Heat or warmth' in the body as a result of eating durian can be remedied by drinking, or rather sipping, water placed in the locules of the husk.
- Eating mangosteen after consuming large amounts of durian is advisable because mangosteen will also reduce the 'warmth' caused by eating durian.
- The sharp spines of the durian husk are a good remedy for rheumatism when scrubbed on the body.
- The durian tree can help maintain family relationships as, while consuming a good quality durian, family members are willing to forgive and forget their disputes.
- The first few fruit off a tree or in a season should be given away, in order to sustain good yields in subsequent years.
- One should not climb the durian tree as it can affect flowering.
- The durian tree must not be clean-weeded or it will be less productive.
- People who consume durian in conjunction with alcohol may become sick or even die.

3

Morphology, Genetics and Cultivars

Durian (*Durio zibethinus*) is one of the best known and most controversial of all South-East Asian fruits. The species' botanical name *zibethinus* is derived from the Italian word 'zibetto' which means strong smelling. The common name 'durian' is derived from the Malay word 'duri' which means thorn. These two names precisely describe the chief characteristics of durian fruit.

Botanical Characteristics

The botanical structure of durian has been studied and documented in detail by many botanists. The major characteristics of *D. zibethinus*, the most popular and domesticated *Durio* species, can be summarised as follows.

Tree size and form

Durian trees have a life span of 80 to 150 years. A full-grown tree can have a trunk diameter of 50 to 120 cm. Under normal conditions, seedling trees can grow to heights of 40 m or more with a canopy width approximately two-thirds the height. The tree has a columnar, buttressed, low-branched trunk and a dense crown. The branches grow from the main trunk in every direction. They may be straight or curved, depending on the cultivar and amount of exposure to sunlight. The bark is a dark reddish-brown, peeling off from many deep longitudinal splits. The larger branches are covered with numerous lenticels. The secondary branches are grooved with a reddish pubescence.

While plants of seedling origin have a straight trunk and a dense, relatively compact branch structure, cultivars which have been vegetatively propagated may have either erect or lax trunks and spreading forms. Clonal cultivars are generally less vigorous than seedling forms.

Root system

The roots do not penetrate deeply into the soil, although the main taproot may extend to several metres. Most of the lateral roots grow near the soil surface. Seedling trees have relatively vigorous roots compared with those of clonal cultivars.

Leaf structure

The leaves are alternately elliptical and lanceolate, 10 to 20 cm long and 4 to 6 cm wide. The leaf blade is shiny, smooth and a dark green on the upper surface. The lower surface is densely covered with silver or gold scales and a layer of stellate hairs. The base of the leaf is obtuse and the apex is abruptly acuminate and fragrant when crushed. The petioles are thick, about 1.5 to 3.0 cm long and 0.15 to 0.35 cm wide. The leaf buds are flat and the young leaves are folded at the midrib.

Flower structure

The large flowers are borne laterally on the pendulous branched inflorescences on older branches. Inflorescences form fascicles of corymbs of 3 to 50 flowers which grow up to 15 cm long. The pedicels are thick, terete, jointed, scaly and 2 to 4 cm long. The epicalyx, the outermost covering of the flower, is entirely closed initially and later splits into two to four segments. The epicalyx segments are concave, deciduous lobes, 1.5 to 2.5 cm in length, greenish-yellow and scaly. The flowers are 5 to 6 cm long and about 2 cm in diameter. The five sepals are 1.5 cm in diameter, 2 to 3 cm high, united below the middle and fall off as a unit.

Five large orange-coloured nectaries are located at the base of the calyx. The five petals are white, light yellow or cream and longer than the calyx. They are deciduous, oblong, spatulate, rounded at the apex, scaly, 3 to 5 cm long and 1 to 4 cm wide. The stamens are numerous and united in five loose bunches located in the middle of the petals. The stamen is 2.5 to 5.0 cm long. Each bundle consists of 4 to 18 different-sized filaments. The anthers are reniform and in clusters of 3 to 12. The ovary is ovoid to oblong in shape, densely covered with silvery scales, contains four to five cells and is 0.5 to 0.7 cm long. The style is terete, thickened at the base, densely pubescent, light salmon-coloured, 3 to 5 cm long and topped by a large yellow or orange stigma.

Durian flowers appear in clusters on the branches or trunk of the tree (Plate 3.1). There are up to 50 flowers per cluster (Somsri, 1987) (Plate 3.2). They are complete or perfect flowers, meaning each flower has a stamen and pistil (Polprasid, 1960) (Figure 3.3). However, when the flowers open, the pistil and the stamen do not necessarily appear at the same time. Normally, the stigma from the pistil will come out first, long before the anthers of the stamen appear and shed their pollen.

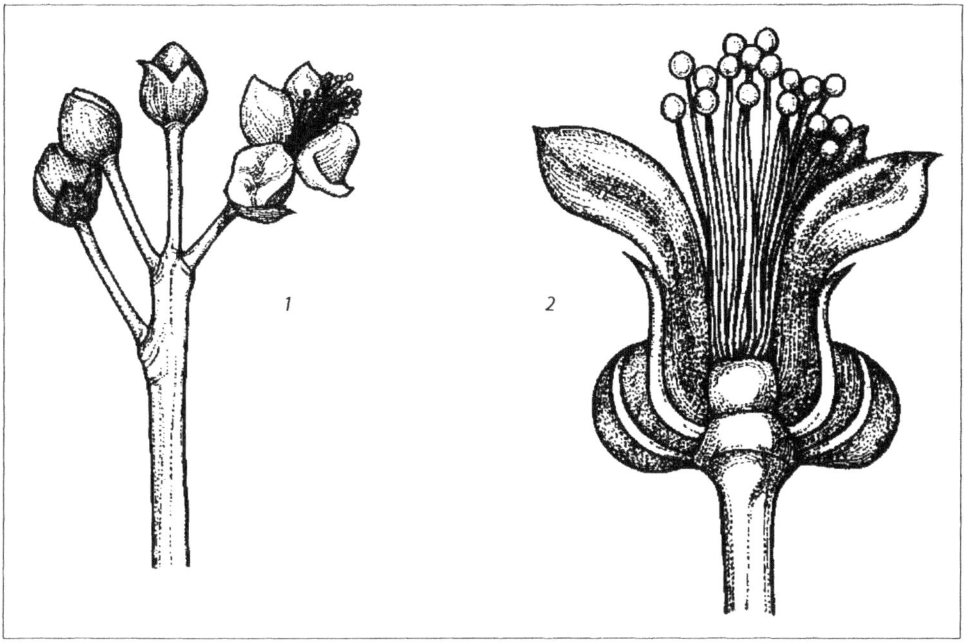

Figure 3.3 *Diagram of durian flower: 1) individual cluster; 2) longitudinal section of individual flower*

Fruit

The fruit, which are located on the underside of the branches, are stalked. They are pendulous, round to oblong, commonly 13 to 16 cm wide and 15 to 25 cm long, but may be up to 35 cm in length (Plate 3.4). The colour varies from olive-green to yellow. The fruit may exceed 3 kg in weight. The fruit peduncle is very thick, 5 to 6 cm long, and 1.0 to 1.2 cm in diameter. The thick rind is covered with broadly pyramidal, coarse, hard, sharp spines or 'duri' (hence the common name 'durian'). The spines, which are about 1.5 cm long and 1.9 cm wide at the base, are slightly angled and abruptly taper to a very hard, sharpened point. Amongst the spines, there are four to five depressions which run from the base to the apex of the fruit. These depressions become cracks which eventually split when the fruit ripens fully.

The fruit is actually a capsule. The four or occasionally three segments are of irregular width and hold up to seven seeds in depressions on each side. Each seed is embedded in a cream or yellow-coloured aril of custard-like consistency (Plate 3.5). The aril is the edible portion of the fruit and varies in colour, texture and thickness between different seedlings or varieties. Seeds are large, about 5 to 7 cm long and 2 to 3 cm wide. They are fairly hard and covered with a thin, light brown skin.

The durian fruit appears to have a similar structure to the jackfruit but is formed from one flower and one ovary, and not from many ovaries as the jackfruit is (Molesworth-Allen, 1965). The rind of the durian fruit is extremely tough and through time has evolved to withstand falling from as high as 40 m without sustaining any significant damage. Durian fruit are food for a great variety of animals, including orangutans, elephants, tigers and even the domestic cat.

Genetics and Cultivars

The durian cultivars grown commercially in the ASEAN countries are derived from wild strains of *D. zibethinus*, which originated in the Malay Peninsula. Existing commercial cultivars arose from chance seedlings, selection by growers and more recently by breeding in government institutions. The natural selection process is most active in Thailand and Malaysia. The names of cultivars grown throughout ASEAN countries differ.

Cultivars in Malaysia

There are at least 28 known *Durio* spp. in Malaysia. The most productive and economical species is *D. zibethinus*. A few other species also produce edible fruit but are of slightly inferior quality so are less widely cultivated. The semi-wild edible *Durio* spp. include *D. oxleyanus*, *D. lowianus*, *D. graveolens*, *D. carinatus*, *D. dulcis* and *D. testudinarium*.

After evaluating the *Durio* spp. for the most desirable characteristics, specialists in Malaysia have developed the first three hybrid durian clones which give better fruit quality and yield, show good disease tolerance, and fruit early (Zainal Abidin, 1992; 1993). Three hybrid clones, MDUR 78, MDUR 79 and MDUR 88, have recently been put on the recommended list of the Malaysian Agricultural Research and Development Institute (MARDI).

D 2

The local name is Dato Nina and it originated in Malacca. The tree is medium to large and erect in form. It flowers regularly but is low yielding. D 2 shows effective resistance to *Phytophthora palmivora* stem canker. The fruit, generally borne on small secondary or tertiary branches, weigh 1.3 to 1.8 kg and are characteristically lop-sided (kidney-shaped). The fruit are difficult to open although the valves are thin with small sharp spines. The aril is thick, copper-yellow in colour and firm. Each locule possesses only a few pulp units, often variable in size. Although the aril quality is excellent, this clone is not popularly planted due to its low yields.

D 10

This clone comes from Gombak, Selangor. The plant is medium in size with a broad canopy. It flowers regularly and is medium to high yielding. This clone is sensitive to *P. palmivora* stem canker. Fruit are borne all over the tree. The fruit is round to oval, weighing 1.0 to 1.7 kg. The fruit does not keep well and tends to split easily. The skin is of moderate thickness and is yellowish-green in colour. The aril, which is considered of good quality, is thick, bright yellow in colour and has a sweet, nutty flavour. Each locule is completely filled with pulp units arranged in single rows.

D 24

This clone originated from Bukit Merah, Perak, and is the leading commercial durian clone in Malaysia. The large and usually vigorous tree has a broad, pyramidal canopy. The trees flower regularly and are very high yielding. By the time the tree is between 10 and 15 years old, it may bear up to 150 fruit per season. Fruit are borne all over the tree but are most plentiful on the larger lower branches. The round or oval shaped fruit are medium to large and weigh 1.0 to 1.8 kg. Each locule contains between one and four large pulp units arranged in single rows. The aril is thick, light yellow in colour and is firm and fine in texture. It is sweet and nutty with a slightly bitter taste. The main disadvantage of D 24 is that it is extremely susceptible to *P. palmivora* stem canker. Fruit of this clone also exhibit some degree of the physiological disorder dubbed 'uneven fruit ripening'.

D 99

This clone originated in Thailand and has become widely planted. The tree is medium-sized, low branching and often has a loose canopy. It flowers and fruits regularly with a tendency towards biannual fruiting. Its fruit bearing ability is very high, bearing up to 130 fruit per season when between 10 and 15 years of age. The clone will tolerate both *P. palmivora* stem canker and a dry environment. It is a good pollinator and is cross-compatible with most other clones. Fruit are borne mainly on the lower branches. The fruit are small to medium in size, weigh 1.0 to 1.5 kg and are almost round in shape with a slight depression at the apical end. They are easily split due to a thin rind. Each locule contains one to four pulp units. The aril is thick, bright yellow, fine in texture and rather moist. It has a sweet nutty taste and a pleasant aroma.

D 123

This clone is commonly known as 'Kani' (or 'Chanee' in Thailand). It is best grown in the dry climate of Malaysia. The fruit are long in shape and hollow at the end. The rind is yellowish-green and easily split. The fruit weigh 2 to 4 kg and contain four to five locules. Each locule consists of one to four large pulp units. The thick aril is yellow and sweet with a rough texture. The aril weight makes up 26% of the total fruit weight. This cultivar, released in 1971, is not resistant to root-rot disease.

D 145

This is a recently recommended Malaysian clone, released in 1991. It is also known as 'Durian Hijau' or 'green durian'. The clone originates from Beserah, Pahang. The tree is medium to large in size and is rather sensitive to dry conditions. It fruits rather irregularly but shows average to high fruit yielding ability. Fruit are borne all over the tree. The fruit are medium to large, weighing 1.3 to 2.0 kg, and are round or oval in shape. The rind is quite thick and dark green in colour and is easily split. Each locule contains one to four large pulp units arranged in single rows. The aril is medium to thick, bright yellow and has a soft texture. It has a pleasant aroma and tastes sweet and nutty. The aril quality is considered very good. This clone is quite resistant to *P. palmivora* stem canker.

D 158

This cultivar is cloned from the Kan Yao cultivar of Thailand. It is most successfully grown in the dry Malaysian climate. The fruit are round with a very long fruit stalk of 15 to 17 cm. The rind is green and easily split. The spines are short. Each fruit weighs 2 to 4 kg with one to four pulp units in each locule. The thick aril is golden-yellow with a sweet taste and a rather fibrous texture. The aril weight consists of 26% of the total fruit weight. This cultivar, released in 1987, is not resistant to canker disease.

D 159

In Thailand and Indonesia, this cultivar is familiarly known as 'Mon Thong', meaning 'golden pillow'. D 159 is also best grown in the dry Malaysian climate. The fruit are cylindrical in shape with a short fruit stalk. The thick rind is yellowish-green and difficult to split. Each fruit weighs 4 to 6 kg and has a golden-yellow aril. The aril is thick, sweet tasting and soft in texture. This cultivar, released in 1987, is not resistant to canker disease.

D 169

This cultivar, originally from Kelantan, was released in 1989. The fruit are long with a thin yellowish-green rind and are easily split. The spines are sharp. Each fruit weighs 1 to 2 kg with one to four pulp units in every locule. The thick aril is golden-yellow, soft in texture and sweet, yet also rather bitter in taste.

MDUR 78

This cultivar is the result of cross-breeding between D 10 and D 24. The plant is relatively small, shady, high yielding and resistant to *P. palmivora* stem canker. Each fruit weighs 1.5 to 1.8 kg and is round or oval in shape. The skin colour is yellow to light green. The aril is thick and each pulp unit is quite large. The aril ranges in colour from orange to yellow, has a fine texture and is sweet and nutty. It keeps well, with a natural shelf life of about 70 hours. This cultivar, released in 1991, is resistant to canker disease.

MDUR 79

This cultivar has also been cross-bred from D 10 and D 24. It is a relatively small plant, fruiting regularly with average yields. It is resistant to *P. palmivora* stem canker. The fruit are round or oval and weigh 1.0 to 1.6 kg. The rind is dark green. The aril is thick and each pulp unit is quite large, orange-yellow in colour with a fine texture and a sweet and nutty flavour. This cultivar, also released in 1991, is resistant to canker disease.

MDUR 88

This is a hybrid from a cross between D 24 and D 10. The plant is medium to large in size and vigorous. It is a high yielding and consistent bearer, with moderate tolerance to *P. palmivora* stem canker. The fruit weigh 1.5 to 2.0 kg, are round to oval in shape and have a yellowish-green rind. The aril is very thick and each pulp unit is large. The aril is golden-yellow in colour, slightly dry, fine in texture and is sweet and nutty. The fruit has a fairly long storage life of between 78 and 86 hours. This cultivar, released in 1992, is resistant to trunk canker disease.

Cultivars in Thailand

Durian has been cultivated in Thailand for over a century and almost 200 cultivar names are recognised. However, only 60 to 80 cultivars have been commercially grown (Bhusiri, 1987), including 21 cultivars in Nonthaburi and the surrounding provinces, which is believed to be the first commercial durian-growing area in Thailand (Kietkong, 1957). Most of the cultivars are from the seedlings of open-pollinated fruit, and the names given to these cultivars were inspired by the characteristics of fruit shape or pulp colour, the persons who selected them, or the places they were found. Many names are similar or only slightly different from one another. To avoid confusion and duplication of cultivars in trading and in breeding programs, methods of classifying them into groups have been developed.

Inthong (1964) suggested three categories of cultivars, depending on the time they take to first bear fruit, as follows:

1) *Early cultivars*: those that bear fruit within four to six years of planting, eg Luang, Kradum and Srichomphoo.
2) *Medium cultivars*: those that bear fruit within six to eight years of planting, eg Kob, Chanee and Thong Yoi.
3) *Late or heavy cultivars*: those that bear fruit after eight years or more, eg E-nak and Kampan.

Bhusiri (1987) suggested a method of classification based on: 1) the appearance of the fruit itself, as in 'Mon Thong' which means 'golden pillow', ie the fruit is shaped like a pillow; 2) the colour or texture of the flesh or pulp, as in 'Kam Pan Daeng' which means 'reddish-yellow pulp'; and 3) the person or agency who selected the cultivar, as in the Maetao cultivar which was named after Mr Kob Maetao. Using these criteria, Bhusiri proposed eight categories as follows:

1) Chanee group	5) Thong Yoi group
2) Kob group	6) Kam Pan group
3) Kan Yao group	7) Luang group
4) Mon Thong group	8) Miscellaneous group

Recently, Hiranpradit *et al.* (1992) proposed a system adopted from the International Board for Plant Genetic Resources (IBPGR) tropical fruit descriptor. Based on leaf characteristics, fruit shape and spine, they proposed six groups:

1) Kob	4) Kam Pan
2) Luang	5) Thong Yoi
3) Kan Yao	6) Miscellaneous

Durian progenies are heterozygous in all attributes. Most durian seedling trees are unlikely to be able to self-pollinate, therefore flowers must be cross-pollinated from other trees in order to ensure fruiting (Valmayor *et al.*, 1965; Malo and Martin, 1979). Cross-pollinating different varieties of trees has created vast genetic diversity. In Thailand, a number of clonal selections from seedling trees have resulted in many new cultivars being introduced.

Durian clones in Thailand differ from one another in terms of: tree form, vigour and leaf size; flowering intensity and frequency; fruit size and shape; rind colour and thickness; aril volume, colour, aroma, texture and flavour; and seed number and size.

There are five original durian varieties in Thailand and the types developed from them are as follows (Vangnai, 1996):

Original variety	*Cultivars developed*
Luang	Luang Khiew, Chanee, Chompoo Sri, Yam Maward
Karaket	Kob Maetoa, Kob Phuang, Kob Lebyiew, Yampleh
Tong sook	Kan Yao, Sri Suwan, Kan Yao Watsak
Thong Yoi	Thong Yoi Chat, Nom Sawan, Sin Samut, Chat Srinak, Chat Ngarm, Chat Srithong, Thoranee Wai
Kam Pan	Kam Pan Luang, Chau Mafai, Kam Pan Srinak

Fruit characteristics of the original durian varieties, as recorded by Bhusiri (1987) and Vangnai (1983), are described as follows:

Luang

The fruit are medium-sized, oval in shape, 20 to 25 cm long by 16 to 20 cm wide and are yellowish-green when mature. They have four to five lobes with slight depressions. The medium-spaced spines are small, the fruit stalk is short and the rind is moderately thick. The yellow aril is medium to thick with a coarse texture and sweet taste. The fruit have five locules with each locule containing one to six pulp units.

Karaket

The fruit are ovoid, medium to large and 25 cm long by 21 cm wide. The rind is greenish-brown and covered with curved, sharp-pointed spines, about 2 cm long. There are five conspicuous lobes with depressions that run from the base to the apex of the fruit. The aril is thick with a fine texture and is dark yellow in colour. It tastes very sweet and nutty, and has a strong odour. The fruit have five locules and each locule contains two to three pulp units.

Tong Sook

The fruit are round, 20 cm long by 19 cm wide with five inconspicuous lobes. The rind is rather thick and covered with dense brownish-green spines of 1.0 to 1.5 cm and a long fruit stalk of 12 to 15 cm. There are five to six locules with three to five pulp units in each locule. The yellow aril is medium to thick with a fine texture. It is sweet and creamy with a mild odour.

Thong Yoi

The fruit are oblong, 27 cm by 21 cm in size with a protrusive apex. They have five conspicuous lobes with depressions which run from the base to the apex. The rind is yellowish-red when mature and covered with large base spines of a greenish-brown colour. The dark yellow aril is thick, sweet and nutty with a strong aroma.

Kam Pan

The fruit are round, medium to large and 20 cm long by 18 cm wide. The fruit shoulder is enlarged, tapering towards the round apex. Each fruit has five lobes and a greenish-brown rind with reddish-brown depressions. The short spines are thick, straight, and slightly angled at the tip. The fruit contain five locules with one to two seeds in each locule. The thick aril is pale yellow to white with a very fine texture. It is very sweet with a strong smell.

The selection of durian varieties in Thailand was based largely on fruit quality. Seedlings were selected by farmers in the early 1880s when they were perpetuated from chance seedlings as clones. As a result of this practice, by the year 1884, there were about 68 good durian clones growing in Nonthaburi and its vicinity. About 200 durian clones are now recognised in Thailand (Vangnai, 1996).

Boonyakom (1955) classified durian clones in Thailand (85 cultivars in all) into six main groups as follows:

1. Kob group consisting of 25 cultivars: Jorm Kob, Kob Maetao, Kob Lebyiew, Kob Boonnak, Kob Takhum, Kob Hemaraj, Kob Chompoo, Kob Thongkon, Kob Khunpaen, Kob Phrawai, Kob Ngoy, Kob Chao Kun, Yan Pleh, Sood Sakhon, Luang Hokepoo, Kob Pai Chum Pon, Chat Thong, Kleeb Samut, Kob Ta-Jam, Kob Yaiplub, Kob Tanoi, Kob Loynam, Jorm Yota, Kob Ta Khiew, Kob Thongkhum.

2. Thong Yoi group consisting of 18 cultivars: Thong Yoi Chat, Nom Sawan, Sin Samut, Chat Ngarm, Phuang Chat, Chat Srinak, Chat Srithong, Chompoo Nuj, Thong Sawat, Thong Yoi Srinak, Thoranee Wai, E-rung, Sri-suk, Kam-mayee, Bangbon, Nok Yib, Daeng Rasamee, Tubtim.

3. Kan Yao group consisting of seven cultivars: Kan Yao, Sri Suwan, Kan Yao Watsak, Kan Yao Namyai, Chompoo Parn, Kan Yao Srinak, Kan Yao Boon Young.

4. Kam Pan group consisting of 11 cultivars: Kampan Khao, Kampan Luang, Kampan Daeng (short leaved), Kampan Daeng (long leaved), Kampan Daeng, Takluay Looklek, Sucreep, Kampan Srinak, Maled Faw, Takay, Chau Mafai.

5. Luang group consisting of 10 cultivars: Luang Khiew, Luang Thong, Luang Loom, Luang Neur-daeng, Leang Sompoy, Laweng, Yum Maward, Chompoo Sri, Chanee, Sri Chompoo.

6. Miscellaneous group consisting of 14 cultivars: Karaket Maetao, Karaket Srinak, Phuang Malai, Krapook Thong, Daeng Tanoi, Thosagun, E-nak, Thong Daeng, Mai Thongyib, Khua Yi, Krapook Thongdee, Kra-toei, Neur-daeng, I-ad.

The popular durian cultivars commercially grown in Thailand at present are: Kan Yao, Chanee, Mon Thong, Kradum Thong, Kob Pikul and Kob Takhum. The first four cultivars have become internationally recognised. In Thailand, Mon Thong and Chanee can be found in all the durian-growing regions, in the southeast, central, south, and north. Meanwhile Kan Yao and Kradum Thong can be found only in the north of Thailand, in the Uttaradit province.

Cultivars in Indonesia

The Ministry of Agriculture has enumerated 17 major varieties of commercial durian in Indonesia. These varieties are classified as superior based on: the colour and appearance of the fruit, large spines, egg or starfruit shape, short stem and a high yield. The most important criteria, however, are that the fruit is fleshy and thick with small rough seeds, a dry smooth texture, purplish-yellow in colour and a sweet yet mildly bitter taste. Two of the varieties were adopted from Thailand. They are Kani (originally Chanee) and Otong (originally Mon Thong), while the 15 other varieties are native to Indonesia. The local names of the 17 varieties are Durian Sukun, Petruk, Sitoking, Otong, Kani, Simas, Sunan, Sihijau, Sijapang, Sawerigading, Lalong, Tamalatea, Tembaga, Siriwig, Bokor, Perwira and Nglumut. Many of these varieties have been cultivated by growers for many centuries. Some existing trees are estimated to be more than 100 years old. Yield per tree ranges from 15 to 800 fruit each season.

Durian Sunan

The fruit of this cultivar are egg-shaped. They have a brownish-green rind and small spines which are widely separated. They weigh 1.5 to 2.5 kg each and have a thin rind which may be easily split. Each fruit has five locules, with 20 to 36 pulp units with flattened seeds and cream-coloured flesh. Each tree at 20 years old produces 200 to 800 fruit per year. The appearance of the fruit is more aesthetically pleasing than other varieties. This cultivar is resistant to root-rot disease and fruit-boring insects. Durian Sunan is originally from Desa Gendol, Boyolali in Central Java where the climate is dry.

Durian Sukun

This cultivar's fruit are long and round with quite a thick yellow rind and small spines which are close together. The fruit are easily split and weigh 2.5 to 3.0 kg. Each fruit has five locules with 5 to 15 pulp units with flattened seeds and yellowish-white flesh. Each full-grown tree produces 100 to 300 fruit per year. This cultivar was released in 1984 as a cultivar resistant to root-rot disease and fruit-boring insects. It is originally from Desa Gempolan, Karanganyar, in Central Java where the climate is also dry.

Durian Sitokong

This cultivar's fruit are long and round with a yellowish-green rind and spines which are closely spaced. The rind is quite thick and difficult to split. Each fruit weighs 2.0 to 2.5 kg and has five to seven locules containing 20 to 30 pulp units with yellow flesh. Each full-grown tree produces up to 200 fruit per year. This cultivar, released in 1984, is originally from Pesar Minggu, South Jakarta, where the rainfall is relatively high.

Durian Simas

The fruit of this cultivar are oblong in shape with a reddish-yellow rind and spines which are closely spaced. The rind is fairly thick and not easily split. Each fruit weighs 1.5 to 2.0 kg, has five to seven locules with 20 to 30 pulp units and a bright yellow aril. Each full-grown tree produces 50 to 200 fruit per year. This cultivar, released in 1984, is also resistant to root-rot disease and fruit-boring insects. Durian Simas is originally from Desa Rancamaya, Bogor, in West Java where the rainfall is very high.

Durian Petruk

The fruit of this cultivar are egg-shaped with a yellowish-green rind and small spines which are closely spaced. The fruit are not easily split. Each fruit weighs about 1.0 to 1.5 kg, has five locules each containing 5 to 10 pulp units and a thick yellow aril. Each full-grown tree produces 50 to 150 fruit per year. This cultivar, released in 1984, is resistant to root-rot disease and fruit-boring insects. It is originally from Desa Randusari, Jepara, in Central Java where the climate is dry.

Durian Otong (Mon Thong)

The fruit of this cultivar are long in shape with a yellowish-green rind and small spines which are closely spaced. The fruit may be easily split. Each fruit weighs 2.0 to 4.0 kg, has 5 to 15 pulp units and a yellow aril. Each full-grown tree produces up to 50 fruit per year. The tree starts bearing at the age of four or five. Unfortunately, this cultivar is not resistant to root-rot disease or fruit-boring insects. Durian Otong is originally from the Nonthaburi province in Central Thailand where there is relatively little rainfall. It is, however, very adaptable and can grow in both wet and dry conditions. This cultivar was released in 1987.

Durian Kani (Chanee)

The fruit are round in shape with a brownish-yellow rind and small spines which are closely spaced. The rind is of moderate thickness and may be easily split. Each fruit weighs 2.0 to 4.0 kg and contains four to six locules and 5 to 18 pulp units. The aril is yellow in colour. Each full-grown tree produces up to 50 fruit per year. The tree starts bearing at the age of four or five. Despite being an early producer, this cultivar is not resistant to root-rot disease and fruit-boring insects. Durian Kani was introduced from Thailand in 1987 and is also known as Durian Bangkok. It is very adaptable and can grow in both wet and dry climates.

Durian Sihijau

The fruit of this cultivar are oblong in shape with a green rind. The rind is 1.2 cm thick with long spines closely spaced. The fruit splits easily. Each fruit weighs 2.0 to 2.5 kg and has five locules each containing 20 to 35 seeds. The yellow aril is sweet with a soft or fibrous texture. Each full-grown tree produces 300 to 400 fruit per year. This cultivar is resistant to root-rot disease and fruit-boring insects, and was released in 1990. Durian Sihijau is originally from Karang Intan, South Kalimantan, where the rainfall is high.

Durian Sijapang

This cultivar's fruit are oblong in shape with a greenish-yellow rind about 0.5 to 1.0 cm thick which is easily split. The spines are well spaced. Each fruit weighs 1.5 to 2.5 kg, and each of its five locules contains 10 to 20 small oval seeds. The aril is light yellow and 1.5 to 2.5 cm thick.

It has a soft texture with a sweet dry taste. Each full-grown tree produces 300 to 600 fruit per year. Durian Sijapang is resistant to root-rot disease and fruit-boring insects. This cultivar was released in 1990 and is originally from Karang Intan, South Kalimantan, where the rainfall is high.

Durian Siriwig

The fruit are egg-shaped with a yellowish-green rind. The rind is 1.2 cm thick, with sharp, long, well-spaced spines, and is easily split. Each fruit weighs 1.5 to 2.0 kg and contains 13 to 15 pulp units. The white aril is sweet, fibreless, of medium thickness and has a strong flavour. A full-grown tree produces 100 fruit per year. This cultivar was released in 1993 and is resistant to root-rot disease. Durian Siriwig is originally from Kampung Sinapeul, Desa Ujung Berung, Kecamatan Sindang Wangi and Majalengka in West Java where the climate is wet.

Durian Bokor

The fruit of this cultivar are cylindrical in shape with a yellowish-green rind, 3 to 5 mm thick. The spines are large and well spaced. Each fruit weighs about 3.9 kg and contains a light yellow aril. The aril is sweet, dry, aromatic and of medium thickness. The texture is soft and fibreless. A full-grown tree produces 150 to 200 fruit per year. This cultivar, released in 1993, is resistant to root-rot disease but sensitive to fruit-boring insects. Durian Bokor is originally from Desa Sukajadi and Majalengka in West Java where the climate is wet.

Durian Perwira

This cultivar's fruit are round with a thin green rind which is easily split. The cone-shaped spines are big and closely spaced. Each fruit weighs 2 to 3 kg and contains a thick yellow aril and 15 to 20 pulp units. The sweet aril is dry with a strong aromatic flavour. A full-grown tree produces between 200 and 300 fruit per year. This cultivar, released in 1993, is resistant to root-rot disease and fruit-boring insects. Durian Perwira is originally from Kampung Sinapeul, Desa Ujung Berung, Kecamatan Sindang Wangi and Majalengka in West Java where the climate is wet.

Durian Sidodol

The fruit are round with a yellowish-green rind, 1.1 cm thick, and are easily split. The cone-shaped spines are closely spaced and blunt. Each fruit weighs 1.5 to 2.5 kg and contains a bright yellow aril. The aril is about 1.2 cm in thickness and tastes sweet. The fruit contain 20 to 25 pulp units and seeds weighing approximately 13.3 g each. A full-grown tree produces 100 to 200 fruit per year. Durian Sidodol, released in 1993, is resistant to root-rot disease and fruit-boring insects. This cultivar is originally from Karang Intan in South Kalimantan where the rainfall is high.

Durian Bantal Mas (golden pillow)

The fruit of this cultivar are triangular in shape with a yellowish-green rind and short spines. Each fruit weighs 3 to 6 kg and contains a pillow-like aril. The yellow aril is sweet, fibreless and soft in texture. The fruit contain 10 to 20 seeds. A full-grown tree produces 400 to 900 fruit per year. The Durian Bantal Mas cultivar is originally from Desa Tanjung Beringin, Kecamatan Kikim, Desa Telekud, Kecamatan Palau Pinang and Kabupaten Lahat in South Sumatra where the rainfall is high.

Durian Sihepi

The fruit are egg-shaped with a brownish-green rind. The rind is 8 to 10 mm thick and is easily split. The small spines are cone-shaped and closely spaced. Each fruit weighs 1.5 to 2.0 kg and contains a thick yellowish-white aril. This aril is sweet, fatty and fairly dry with an aromatic flavour and a soft texture. Each fruit contains 5 to 15 pulp units in each of its five locules. Durian Sihepi, originally called 'Hepe' (meaning 'flat seeds' in the Sundanese language), contains partially flat seeds with only two or three normal seeds. A full-grown tree produces 150 to 250 fruit per year. It is resistant to root-rot disease and fruit-boring insects. Durian Sihepi is originally from Desa Bendungan, Kecamatan Jonggol and Kabupaten Bogor in West Java where the climate is wet.

Durian Matahari

The fruit of this cultivar are cylindrical in shape with a brownish-green rind. The rind is 5 to 10 mm thick and is easily split. The large spines are sharp, crooked and well spaced. Each fruit weighs 2.0 to 3.5 kg and has a thick, bright yellow aril. The sweet flesh is dry and aromatic with a soft fibrous texture. The fruit contains 10 to 20 pulp units located in each of the five locules and 5 to 10 normal seeds. A full-grown tree produces up to 200 fruit per year. Durian Matahari is originally from Desa Cimahpar, Kecamatan Kedung Halang and Bogor where the climate is wet.

Durian Aspar

The fruit are cylindrical in shape with a brownish-green rind. The rind is 1.0 to 1.5 cm thick and is easily split. The short spines are cone-shaped and well spaced. The fruit weighs 6 to 8 kg and contains a light yellow aril. The aril is sweet-tasting, soft, dry and aromatic. It is 1.5 to 2.5 cm thick and weighs 30% to 40% of the total fruit weight. Each fruit contains 18 to 22 pulp units located in each of the five locules and seeds weighing 25 g each. A full-grown tree produces between 150 and 200 fruit per year. Durian Aspar, released in 1995, is resistant to root-rot disease and fruit-boring insects. It is originally from Desa Palaman Mabah, West Kalimantan, where the rainfall is relatively high.

Durian Sawah Mas

This cultivar's fruit are egg-shaped. The green rind is 1.0 to 1.3 cm thick and is easily split. The spines are cone-shaped and well spaced. Each fruit weighs 2.5 to 4.0 kg and contains a yellow aril. The aril is sweet-tasting, dry and soft and weighs 1.3 to 1.8 kg. The fruit contain between 12 and 16 pulp units in the five locules with 12 to 14 seeds. A full-grown tree produces 150 to 250 fruit per year. Durian Sawah Mas, released in 1995, is resistant to root-rot disease and fruit-boring insects. It is originally from Desa Mabah in West Kalimantan where the rainfall is quite high.

Durian Raja Mabah

The fruit are oval or oblong in shape with a green rind. The rind is 1.0 to 1.3 mm thick and is easily split. The spines are cone-shaped and well spaced. Each fruit weighs 2.5 to 5.0 kg and contains a yellow aril. The sweet flesh is dry and aromatic; it is 1.5 to 2.0 cm thick and weighs 1.5 to 2.5 kg. The fruit contains 14 to 18 pulp units in each of the five locules with 10 to 15 seeds which weigh 23 g each. A full-grown tree produces 150 to 200 fruit per year. Durian Raja Mabah cultivar is resistant to root-rot disease and fruit-boring insects. It is originally from Desa Mabah in West Kalimantan.

Durian Kalapet

The fruit of this cultivar are oval-shaped with a yellowish-green rind, 1.0 to 1.3 cm thick. The rind is easily split. The spines are cone-shaped and well spaced. Each fruit weighs 2.0 to 3.5 kg with a light yellow aril, 1.5 to 2.5 cm thick. The aril weighs 0.6 to 1.0 kg and is sweet-tasting, dry and soft. Each fruit contains 15 to 18 pulp units in the five locules. The seeds are flat, similar to that of Durian Sihepi. A full-grown tree produces 150 to 200 fruit per year. Durian Kalapet is resistant to root-rot diesease and fruit-boring insects. It is originally from Desa Kayu Tanam in West Kalimantan.

Durian Lai Mansau

This is the only durian variety from *D. kulejensis* but it can be classed as cultivated durian because it is cultivated in some locations. The fruit are round in shape with a yellow rind, 0.9 cm thick. The rind is easily split. The small spines are cone-shaped, sharp and closely spaced. Each fruit weighs 0.9 to 1.5 kg and contains a dark red aril. The aril is sweet, dry and soft, has no aroma and is 0.5 to 1.0 cm thick and 460 g in weight. Each fruit contains 14 to 17 pulp units located in five locules with between 14 and 16 seeds. A full-grown tree produces 200 to 350 fruit per year. Durian Lai Mansau is resistant to root-rot disease and fruit-boring insects. This cultivar is originally from Nanga Pinoh in West Kalimantan where the rainfall is high.

Cultivars in the Philippines

There are six durian cultivars recommended for planting in the Philippines:

DES 806

The fruit of this locally selected cultivar are ellipsoidal in shape. They weigh 2 to 4 kg and are greenish-yellow in colour when ripe. The aril is yellow, sweet, glutinous and slightly bitter to taste. About 25% of the fruit is edible.

DES 916

This is another local selection. The fruit are generally ellipsoidal in shape, weigh 2 to 4 kg and are a greenish-brown colour when ripe. The aril is yellow with a sweet taste and a glutinous texture. About 25% of the fruit is edible.

Chanee

This cultivar was introduced from Thailand. The fruit are ovoid with a broad and obtuse tip. They weigh 2 to 5 kg and are greenish-brown in colour. The aril is golden-yellow and sweet. About 32% of the fruit is edible.

Mon Thong

This cultivar was also introduced from Thailand. The fruit are slightly elongated with a pronounced beak at the base. They weigh 2 to 5 kg and are yellowish-brown in colour. The aril is creamy-yellow and sweet. About 30% of the fruit is edible.

Umali

This cultivar was developed from a seedling introduced from Thailand by the late Dean Umali of the University of the Philippines at Los Banos (UPLB). The fruit are generally round to elongated in shape. They weigh 2 to 3 kg and are yellowish-brown in colour. The aril is golden yellow and sweet. About 32% of the fruit is edible.

CA 3266

This cultivar was introduced from Indonesia. The fruit are round in shape, weigh 1.5 to 2.5 kg and are greenish-yellow when ripe. The aril is light yellow and sweet. About 25% of the fruit is edible.

Cultivars in Brunei Darussalam

Few known *D. zibethinus* clones originated in Brunei Darussalam but some have been imported from Malaysia and Thailand (Department of Agriculture, Brunei Darussalam, 1988). However, six local *Durio* spp. have been identified (Tinggal, 1992): *D. graveolens* Becc.; *D. testudinarium* Becc.; *D. oxleyanus* Griffith.; *D. kutejensis* Becc.; *D. dulcis*; and Durian Suluk which is probably a natural hybrid between *D. zibethinus* and *D. graveolens*.

The Department of Agriculture, Brunei Darussalam, has collected and registered the best species. Clones BD 1, BD 22 and BD 62 are included in the initial collections of these natural hybrids, as they have exceptionally good aril quality.

Cultivars in Singapore

The three durian cultivars grown in Singapore are described as follows:

H.C. Tan No. 2

This locally selected cultivar has pear-shaped fruit. Each fruit weighs 1 to 2 kg and is light green in colour with medium-sized spines. The aril is creamy and thin in texture, sweet, yet with a slightly bitter taste. This cultivar occasionally produces seedless fruit.

H.C. Lim

The fruit of this cultivar are long and oval. They weigh 1 to 2 kg and are brown with medium-sized spines when ripe. The fruit contains a thin, pink, sweet-tasting aril. The aril is not fibrous and the seeds are generally small.

Lim Keng Meng

These fruit are round in shape. They weigh 1 to 2 kg and are brown with medium-sized spines. The aril is yellow in colour, medium-sized and creamy with a bitter taste.

Cultivar adaptability

One of the prerequisites for successful commercial durian growing is the selection of suitable cultivars. Other elements contributing to this success are climate, irrigation, fertilisation, and pest and disease control. These factors are the reason behind the many different durian cultivars in Indonesia and Malaysia; the agro-climates of durian-growing centres in Indonesia and Malaysia are extremely varied

Why then does Thailand, with some 200 cultivars (60 to 80 of which are suitable to be expanded for commercial sale) concentrate on only four leading cultivars? The reason is that farmers consider these four particular cultivars to be of superior quality. The most important factor supporting these considerations is adaptability, such as that displayed by the cultivars Mon Thong and Chanee.

Mon Thong and Chanee trees can be found in all centres of durian cultivation in Thailand, ie in central Thailand (Nonthaburi province and its environs), eastern Thailand (Rayong, Chantaburi, Trat, Prachin Buri provinces and their environs), and northern Thailand (Uttaradit province and its environs). The agro-climate in central and eastern Thailand is dry, so Mon Thong and Chanee cultivars are particularly dominant in those regions. Kan Yao and Kradum Thong cultivars are widely grown in Uttaradit, a province situated between Bangkok and Chiangmai. The agro-climate of Uttaradit is relatively more humid than that of central and eastern Thailand.

Durian farmers in Malaysia also commonly experience a humid climate. The widest centre of durian cultivation is on the east coast of Peninsular Malaysia, followed by the west coast areas of Sabah and Sarawak. In the east coast regions, such as the areas surrounding Perak and Tapah, farmers generally grow cultivar D 148. D 158, D 159, and D 169 are more suitable for regions like Kedah, Palau Pinang, Kelantan and Terengganu where the climate is relatively dry.

Cultivar D 168 is suited to conditions in Johore whereas cultivars D 99 and D 123 (being highly adaptable) can be grown in nearly all centres of durian cultivation. D 24, which originates from Bukit Merah where there is a high rainfall, can also be successfully grown in a dry climate. Similarly, cultivars D 99 and D 123, introduced from Thailand, can be grown in either the dry or wet regions of Malaysia.

4

Phenological Development

Flowering

There are up to 30 flowers in each durian flower cluster but only two or three fruit are harvested from each cluster. Individual flowers from the same flush do not develop at the same rate. About 40 to 50 days after flower buds become visible, organogenesis is completed. The flower swells and white petals appear. This phenomenon usually occurs in the morning at about 0800 to 0900 hours and the pistil is ready for pollination in the afternoon at around 1400 to 1500 hours. This is seen in the extension of stigma above the flowers. Full bloom occurs at around 1600 to 1700 hours and the pollen becomes active at 1800 to 1900 hours. The daily activities of various floral parts related to the pollination of four durian cultivars in Thailand are shown in Table 4.1.

Table 4.1: Activities of various floral parts of four durian cultivars

Cultivars	*White petal stage*	*Pistil receptive to pollination*	*Full bloom*	*Pollen becomes active*
Luang	*0900 h*	*1330 h*	*1645 h*	*1815–1900 h*
Kradum	*0900 h*	*1330 h*	*1600 h*	*1745–1800 h*
Khiew Sa-ard	*0845 h*	*1330 h*	*1600 h*	*1745–1800 h*
Chanee	*0900 h*	*1400 h*	*1630 h*	*1830–1900 h*

Pollination

With some Thai cultivars, Polprasid (1960) found that the stigma started to be receptive at about 1330 hours. Full bloom started at about 1600 to 1645 hours. The anthers shed their pollen from about 1745 to 1900 hours. At 1930 hours, the ability of the pollen to germinate was very low and the few pollen grains that succeeded in germinating did so between 2030 and 2130 hours. Thus self-pollination is rare because when the pistil opens and the stigma is receptive, the stamen is still inside the flower. By the time the pollen is active, the stigma is no longer receptive. Somsri (1987) found that under natural and self-pollinated conditions, the rate of fruit setting was 0% to 6% in Chanee cultivars and 21% in Kan Yao cultivars. Cross-pollination was found to increase fruit set tremendously in these two cultivars. After two weeks of hand-pollination using pollen from different cultivars, the percentage of fruit set was 30% to 64% in Chanee cultivars and 87% to 90% in Kan Yao cultivars.

Valmayor *et al.* (1965) also reported that pollination is problematic in durian cultivars in the Philippines. Since flower anthesis takes place at night, and the flower is large and produces a musky odour, pollination under natural conditions is carried out by bats and nocturnal insects which feed on the nectar of the flower. However, even under natural conditions, the opportunity for pollination is limited because the anthers shed pollen at 1900 hours and fall off before midnight. Cross-pollination or hand-pollination was found to improve both fruit shape and fruit size in the Chanee cultivar (Somsri, 1987). In Kan Yao cultivars, cross-pollination or hand-pollination improved fruit shape but the fruit size was slightly smaller than fruit resulting from natural or self-pollination. This was attributed to heavy fruit set (Somsri, 1987).

Experiments in Thailand on pollination of the Chanee cultivar showed that at 14 days after pollination, the flowers that were left to pollinate naturally had a low fruit set of 0% to 6%. The same fruit set rate applied to flowers that were pollinated with pollen from the same tree or from different trees of the same cultivar. However, the percentage of fruit set rose from 30.1% to 63.6% when they were pollinated with pollen from different cultivars. The highest percentage of fruit set (63%) was achieved using Mon Thong pollen (Somsri, 1987).

At 114 days after pollination, a similar trend was seen; a low rate of fruit set was found in naturally pollinated flowers and those which received pollen from the same cultivar, whereas a high fruit set percentage was found in cross-pollinated flowers (Table 4.2).

An interesting finding of the Chanee experiments concerned the shape of developed fruit. Chanee fruit that were naturally pollinated and those pollinated with Chanee pollens, either from the same tree or from different trees, were irregular in shape, whereas fruit resulting from the pollen of other cultivars developed into a regular shape with a well-developed aril in every locule and normal seeds. However, neither the quality nor the taste of the aril of fruit pollinated with pollen from various sources differed from that of naturally or self-pollinated fruit.

In another study using flowers from the Kan Yao cultivar, it was shown that at 14 days after pollination, the naturally pollinated flowers had a fruit set rate of 21.2%, whereas those

Table 4.2: Percentage of fruit set at 14, 109 and 114 days after pollination of Chanee flowers with pollen from different sources (Somsri, 1987)

Pollen source	Number of flowers pollinated	Percentage of fruit set: 14 days after pollination	109 days after pollination	114 days after pollination	Number of fruit at 114 days after pollination
Natural pollination, flowers unthinned	683	4.1	0.6	0.6	4
Natural pollination, flowers thinned	190	0	0	0	0
Chanee from the same tree	197	6.0	1.0	0.5	1
Chanee from different trees	190	5.6	0.5	0.5	1
Mon Thong	187	63.6	25.7	25.7	48
Kan Yao	187	58.0	27.3	27.3	51
Kob Pikul Thong	186	52.6	12.0	11.8	22
Kra Tuey	197	30.1	5.1	4.6	9

Table 4.3: Percentage of fruit set at 14 and 109 days after pollination of Kan Yao flowers with pollen from three different sources (Somsri, 1987)

Pollen source	Number of flowers pollinated	Percentage of fruit set: 14 days	109 days	Number of fruit at 109 days
Naturally pollinated	323	21.2	0	0
Chanee	109	87.1	31.2	34
Mon Thong	114	89.7	30.7	35

receiving pollens from Chanee and Mon Thong cultivars had a fruit set rate of 87.1% and 89.7% respectively (Table 4.3).

The results of these studies indicate that cross-pollination can help to solve the problem of fruit setting in durian to a certain extent. They also show that the degree of success in pollination varies according to variety, Chanee being more responsive to cross-pollination than other cultivars.

In artificial pollination, the flowers may be pollinated at two stages: while they are open (Type A flowers); or when they are still closed but their epicalyx has already split (Type B flowers) (Coronel, 1983). Type B flowers must be emasculated to free the stigma. The flowers to be used as a pollen source are collected a few hours before they open and their pedicels are immersed in water to keep the flowers fresh. To extend their use, they may be kept in a refrigerator. In pollination, the anther lobes are brushed against the stigma. Results of cross-pollination studies done in the Philippines are shown in Table 4.4. The data showed that pollination of Type B flowers gave a much higher rate of fruit set (87% to 90%) than that of Type A flowers (Coronel, 1966).

Table 4.4: Percentage of fruit set when durian is cross-pollinated (Valmayor et al., 1965)

	Percentage of fruit set	
Cross *	1961	1962
1 x 2[a]	31.0	70.4
1 x 2[b]	–	88.2
1 x 2[c]	–	87.3
1 x 3[a]	33.3	60.0
1 x 3[c]	–	90.0
2 x 1[a]	0	53.3
2 x 3[a]	–	75.0

* Numbers 1, 2 and 3 refer to tree used
[a] Fully opened flowers pollinated
[b] Unopened flowers pollinated at normal anthesis time
[c] Unopened flowers pollinated in the morning before anthesis

Inter-Cultivar Harmony

For successful production, farmers are encouraged to grow a variety of cultivars in their orchards as each durian cultivar has a different capacity for self-pollination. Generally, durian has a low self-pollinating capacity and it needs cross-pollination to achieve good production. Mon Thong is known to have a high capacity for self-pollinating. Therefore, the selection of cultivars that are compatible with one another is recommended when establishing a durian orchard. Malaysia seems to be more advanced than Thailand and Indonesia in researching inter-cultivar harmony.

In Malaysia, research has found that one cultivar may be able to be paired only with another particular cultivar. For example, research carried out by M. Zainal Abidin at Sime Darby Farm in Milai, Negeri Sembilan, has shown that D 24 will have a higher production rate if grown together with D 99 and D 98 or D 144; the recommended combination being 50% D 24, 30% D 99 and 20% D 98 or 20% D 144. The highest quality fruit was produced by D 24, followed by D 99, while fruit from D 98 and D 144 was of average quality. Furthermore, the productivity of D 24 increased. This combination also prolonged the production time on a farm. Cultivar D 99 could be harvested early in the season, ie within 90 to 100 days of pollination; cultivar D 24 could be harvested in the middle of the season, ie within 105 to 115 days of pollination; and cultivar D 98 and D 144 could be harvested at the end of the season, ie within 120 to 130 days of pollination.

Fruit Growth and Development

Studies on durian fruit growth have been conducted in the Philippines (Namuco, 1978) and Thailand (Somsri, 1988; Subhadrabandhu and Shoda, 1997). If pollination is not successful, as in self-pollinated flowers, the pistil shrivels and falls off about four days after anthesis. If pollination is successful (Plate 4.1), the colour of the young fruit changes from light brown to light green and within a week, the ovary starts to enlarge. The ovary has five locules, each locule containing seven to nine ovules. However, not all the ovules develop and a mature fruit contains an average of only 13 mature seeds. Accompanying the enlargement of the ovary is a considerable increase in the size of the fruit peduncle. Two weeks after pollination, the spines start to develop into fine soft protruberances.

The pattern of durian fruit growth, whether expressed in fruit length, fruit width or fruit volume, follows a simple sigmoid curve (Figure 4.2). By the second week after pollination, there is little difference between fruit length and diameter, and this condition continues until the fourth week. From the fifth week onward, there is a more rapid increase in length compared with diameter. Growth is very rapid until the thirteenth week and then it slows down until the sixteenth week when the fruit starts to fall, having reached maturity. In the Chanee cultivar, during the first ten weeks after pollination, the length and width of naturally or self-pollinated fruit are slightly greater than that of cross-pollinated fruit. From the tenth week to the fifteenth week, which is close to harvest time for the Chanee cultivar, fruit which have been cross-pollinated using Mon Thong pollen show a greater increase in fruit length and fruit width than those developed from natural pollination (Somsri, 1988). In the Kan Yao cultivar, fruit developed either from Chanee or from Mon Thong pollen show similar increases in length and width throughout their growth periods. During the tenth and eleventh weeks, they also show better growth than those fruit developed from natural pollination.

Fruit growth periods vary according to the cultivar. Some cultivars take a longer time to attain full size. D 144, a cultivar from Malaysia, takes a long time to reach maturity. This cultivar shows a rapid increase in fruit length and width from initial fruit set up to about 90 days. Fruit development subsequently decelerates and there seems to be no increase in

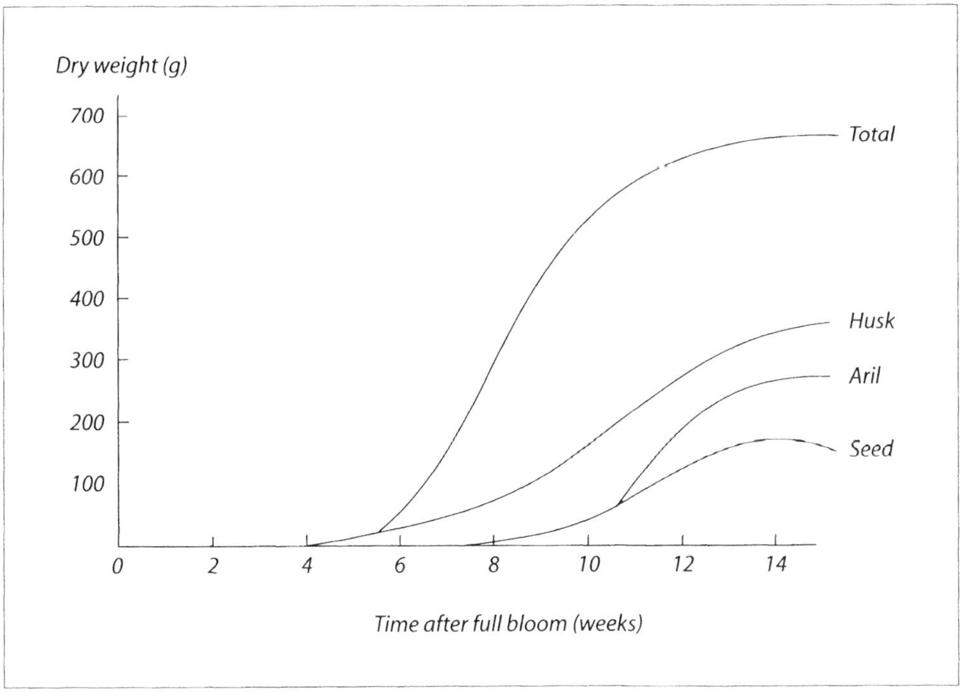

Figure 4.2 *Growth of Chanee cultivar fruit after fruit set (Jantrajoo, S., personal communication)*

fruit size after 100 days. The ripe fruit begin to drop naturally about 115 days after fruit set. By comparison, the growth period of D 99 durian fruit is much shorter. D 99 fruit attain maximum size about 80 days after fruit set and the ripe fruit fall naturally to the ground from 90 days onward after fruit set.

Flowering and Fruiting Seasons

In areas 10° to 18° north or south of the equator, differentiation and flowering normally occur during or near the end of the dry spring months, and fruit development and harvest occur during the wet summer period. Within 10° of the equator the pattern is more complex and in several areas in Malaysia and northern Indonesia, there are two crops a year with peak harvests in December and June. The June crop may be as large as the December crop which is the normal crop in these areas. In any one year, some areas may not produce a crop in either June or December. In Java, where there are more defined wet and dry seasons than on the west coast of Peninsular Malaysia, there is only one cropping period each year. Northern Queensland has a projected harvest season of March to May which coincides with the harvesting time in some areas in Thailand. Table 4.5 depicts the harvest periods for durian in various countries.

Table 4.5: Harvesting periods for durian (Watson, 1978)

Location	*Jan.*	*Feb.*	*Mar.*	*Apr.*	*May*	*Jun.*	*Jul.*	*Aug.*	*Sep.*	*Oct.*	*Nov.*	*Dec.*
India 18° to 8°N						····—	—	—	—			
Myanmar 18° to 10°N					····—	—	—	—	—····			
Thailand 18° to 6°N			····—	—	—	—	—	—	—			
Philippines (Luzon) 15°to 12.5°N								····—	—	—		
Philippines (Mindanao) 9° to 6°N	—	—									—	—
Malaysia (Peninsular) 7° to 2°N	—	—				····—	—	—	—		—	—
Indonesia (Java) 6° to 8°S	—	—								····—	—	—
Papua New Guinea (Bougainville) 5° to 7°S				····—	—	—	—					
Queensland 10° to 18°S			····—	—	—							

Main harvest ——— *Extended harvest (depending on variety and location)* ····———

Temperature and Altitude

The ideal temperature range for good durian growth is 27°C to 30°C, with humidity of 75% to 80%, similar to conditions in a tropical forest. Durian is strictly a tropical plant and can be grown between latitudes 18° north and 18° south of the equator. It is cultivated as far as 18° north in Myanmar, Thailand, Vietnam and India but commercial production is more common and more successful within 12° of the equator.

Compared with most other 'truly tropical' fruits, durian is surprisingly tolerant of low mean temperatures. Terra (1952) lists five centres in Java where durian is grown between 400 and 600 m above sea level, and one special centre, Puspo in East Java, at an altitude of 650 m. However, although relatively cool at this altitude (mean annual temperature 23°C), there are no great extremes as there are at, say, 15° to 17° north or south of the equator, particularly in areas with continental weather influences (Table 4.6). At Kamerunga Horticultural Research Station in northern Australia, the vegetation of durian trees grows actively during all months except June, July and August, which suggests that vegetative growth is limited at mean monthly temperatures below 22°C. In such locations, extension growth halts during the coolest months but flowering and fruiting appear to be more prolific than in non-seasonal equatorial climates. Occasional temperatures of below 10°C appear to cause premature leaf drop (Watson, 1984). At the other extreme, durian trees appear to tolerate temperatures of up to 46°C in some areas of Thailand and India.

Table 4.6: Temperature data for durian-growing centres

Area	*Latitude*	*Altitude m.a.s.l.*	*Mean Annual min. temp. (°C)*	*Mean Annual max. temp. (°C)*	*Mean Annual normal temp. (°C)*	*Mean Min. Coldest monthly temp (°C)*
Cairns (N.Queensland)	*17°S*	*2*	*20.7*	*28.9*	*24.5*	*16.7*
Burliar (India)	*12°N*	*730*	*18.0*	*29.5*	*24.0*	*15.5*
Tambora (Java, Indonesia)	*8°S*	*500*	*20.0*	*25.0*	*22.5*	*17.0*
Chantaburi (Thailand)	*12°N*	*3*	*23.2*	*31.6*	*26.8*	*12.0*
Rayong (Thailand)	*12°N*	*3*	*24.8*	*31.7*	*28.0*	*13.8*

In South-East Asia, durian grows well in areas at an altitude of not more than 600 m above sea level. Although Soeseno (1968) and Anon (1975) recommend the planting of durian up to an altitude of 800 m in Indonesia, numerous authors indicate that durian grows and crops best at medium to low altitudes in the tropics, the ideal range being 50 to 500 m above sea level. If it is grown in the higher altitudes, the quality of the fruit will be affected. At an altitude of 900 m, the durian tree will not produce any fruit at all.

Precipitation

Many varieties of durian are found growing naturally in areas of high precipitation, ie rainfall over 100 mm per month. The ideal rainfall is in the range of 1 500 to 2 500 mm per year, with nine to eleven wet months. However, durian trees can still grow in areas with only seven or eight wet months. Light intensity must be high to stimulate the process of flowering and fruit maturation.

In most Asian production areas, mean yearly rainfall exceeds 2 000 mm but the better production sites have developed in areas where rainfall is well-distributed throughout the year and mean yearly totals exceed 3 000 mm. Malo and Martin (1979) suggest that established durian trees cannot withstand more than three months drought without suffering irreversible damage. However, there is no doubt that with irrigation, drier zones can produce good crops. In some areas, principally at the base of hills and mountains, there is sufficient year round seepage or a high enough water table to support durian production even when the precipitation for that area is fairly low (Terra, 1952).

At altitudes exceeding 200 m in Indonesia, rainfall is generally more evenly distributed throughout the year, and this supports the fairly extensive cultivation of durian in the higher zones. In coastal north Queensland, durian trees grow better on land which has a high water table but is not swampy, or where supplementary irrigation is provided in the spring and early summer months up to the start of the wet season (Watson, 1978).

Soil Requirements

The durian tree can be grown successfully in a wide range of soils but optimum growth and fruiting occurs on rich, deep, well-drained, sandy loams or clay loams which are rich in organic matter (Coronel, 1977). The most commonly used types of soil are alluvial soils derived from basaltic rocks but crops established on red lateritic soils are normally successful when adequately fertilised with a combination of organic and inorganic fertilisers (Yaacob and Subhadrabandhu, 1995).

The durian tree has shallow roots (+1 m) but a long primary root, and therefore needs land with deep sediment/sandstone. Good drainage is essential as the tree is easily prone to root rot disease if the soil is not well-drained. Thus clay soils with poor drainage should be avoided. Ground water should not be too shallow; ideally 2 m beneath the soil surface. Topsoil should be deep to encourage root development for good growth.

In many areas of South-East Asia, durian thrives well on hillsides with moderate slopes, particularly in those areas which have recently been cleared of forest. As durian is susceptible to poor drainage, the ideal location for durian growth is flat or slightly sloping land (not more than 35°). Soil moisture should be regularly maintained at a high level. In the dry season, the subsoil should retain sufficient moisture to sustain floral development – a requirement similar to that of many tropical fruit trees which flower during the dry season. Verheij (personal communication) suggests that the dry period is essential for the drying out of the topsoil which induces a reduction in vegetative growth. The use of an organic mulch during the dry period, as a means of conserving soil moisture, is generally recommended.

A pH range of 4.5 to 6.5 is considered appropriate; at higher pH levels, iron and zinc deficiencies are commonly observed, leading to a yellowing and chlorosis of the leaves (Watson, 1984). The response of various cultivars to differing pH levels is not known.

Prabhadpongse Vechacheva (personal communication), a leading durian grower in the Klung district of Chanthaburi province in Thailand, believes that organic animal manures should be avoided for durian trees, especially in the wet season, since they are conducive to *Phytophthora palmivora* infestation. In nursery conditions in north Queensland, seedlings sustain markedly more vigorous growth when potted into light sandy soils rather than soils with high clay content.

Soil condition is important to durian growth and it can be manipulated by land cultivation and large-scale fertilisation. Although durian can adapt to various kinds of soil, the most suitable is loam or sandy loam mostly found on mountain slopes.

Part 2

Pre-Harvest Management

5

Propagation

Propagation techniques used in durian cultivation include growing from seed and vegetative propagation. Several kinds of vegetative propagation techniques have been developed, eg budding, grafting and layering, and some are more popular than others in the main durian-producing countries. The possibility of multiplication by tissue culture is still being explored.

Seed Selection and Preparation

The seeds should be collected from fully mature fruit. After extraction from the aril, the seeds should be thoroughly cleaned with water to remove any remaining flesh and any undeveloped or misshapen seeds should be discarded. They should be air-dried and coated with fungicide before planting. Durian seeds lose their viability very rapidly (Mortensen and Bullard, 1964) and need to be planted without delay. Seeds may be shipped to distant places, provided they are packed in a moist sterilised medium, but their viability is greatly reduced.

Seed Germination

Seeds for germination can be handled in many ways. They may be sown in a seedbed with a medium of fine sand, or a mixture of sand and garden soil. The seeds should be spaced about 4 to 6 cm apart and planted at a depth of 1 cm. The germination medium should always be kept moist and the seedbed should be shaded (Plate 5.1). As durian seeds are large, they may be sown directly in individual polythene bags of appropriate size (10 x 20 cm or 15 x 30 cm) containing rich, porous soil. In Malaysia, seeds are sown in a damp shady place and planted individually in containers as the radicle emerges. This practice guarantees a higher 'strike rate'. Germination time varies from 7 to 17 days (Mortensen and Bullard, 1964; Padolina, 1931).

In Thailand, many growers germinate durian seeds for rootstock. The seeds are germinated in the above manner until the seedlings are about 20 cm long, or when the seed coat is shed but the leaf is not fully expanded. The seedlings are lifted, the roots trimmed, and they are planted in plastic bags of 8 x 12 cm with two to four drainage holes, in a medium of coconut coir and soil. They are well-watered and kept under shade for about one week, and can be used for inarching (or approach grafting) with the desired variety. A similar method is applied for preparing seedling stock for grafting. In this case the seedlings are lifted early in their germination and kept in plastic bags until the stems are approximately 20 to 30 cm long, or roots have developed.

Seedling Care

Seedlings raised in beds are transplanted into containers (or polythene bags) before the first leaves open, to ensure much faster recovery. If transplanting is delayed, the leaves should be allowed to mature, but must be pruned to reduce transpiration. After transplanting, the seedlings should be watered immediately and placed in the shade. They may be gradually exposed to a moderate amount of light. For faster growth, seedlings should be watered regularly and the soil around the stem cultivated to increase aeration. The seedlings may be given a small amount of nitrogen fertiliser to further enhance growth. They are ready for asexual propagation when they are six to eight months old, or when their stems have attained pencil-size diameter.

Propagation Techniques

Traditionally, durian was propagated from seed, hence the huge variation in tree performance and fruit form. A superior durian tree quickly gains a reputation in its neighbourhood and its fruit are sought out, the seeds sown in new places, increasing the likelihood of variation due to ongoing selection. The result of this practice has meant that some improvements have been made to durian clones. Presently, durian growers use asexual reproduction, which has developed to include many methods. In Malaysia, Indonesia and the Philippines, the modified bud graft, wedge, whip and approach grafts are common. In Thailand, durian growers use various propagation techniques which include the following:

Growing from seed

This type of propagation is necessary when growers want to produce new clones (Plate 5.1). In many cases, plants grown from seed are used as rootstock. Problems include the rapid loss of viability and susceptibility to root disease caused by *Phytophthora palmivora*. In areas where this disease occurs, resistant rootstock should be used. Research on the tolerance of various durian species to this *P. palmivora* root rot is still in progress in Thailand.

Air layering or marcotting

This method is popular with growers in remote rural areas. Air layering or marcotting involves inducing the development of roots on a selected stem while it is still attached to the parent tree. A ball of earth, sawdust, or coconut coir is bound around a branch and kept moist (Figure 5.2). The stem then develops a root system. After a period of two to three months the root system will develop and this new air-layered plant is ready to be removed from the parent plant. The seedling should be planted in a polythene bag and slowly hardened in the shade before being planted out. The advantage of air layering is the high degree of success with which the plant will develop roots. It is a relatively simple process and requires less skill, effort and equipment than many other methods. It also retains the characteristics of the parent tree. Disadvantages include susceptibility to root rot disease, a tendency to a short life span and the risk of being blown over due to the lack of a tap root.

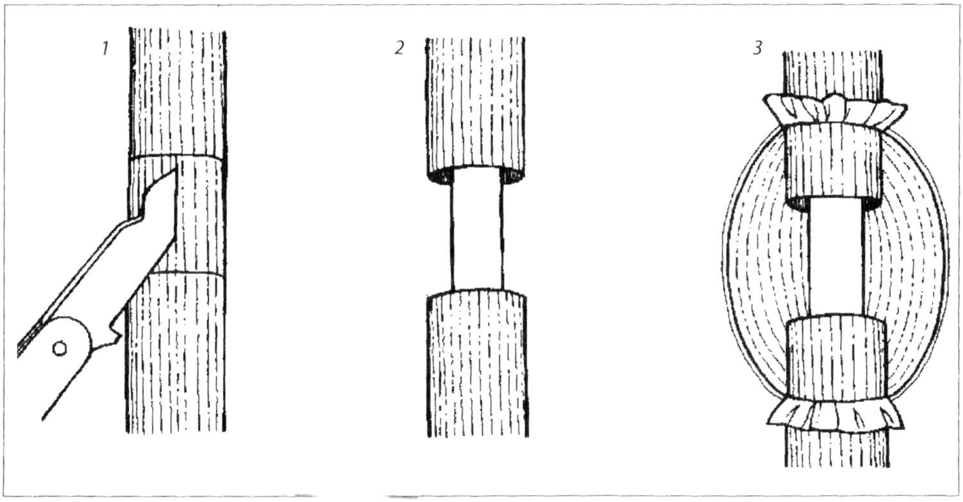

Figure 5.2 *Air layering of durian: 1) girdling the stem to peel off the bark; 2) after the bark is peeled, 3) wrapped with moist medium to initiate roots*

Budding

Generally, growers in Thailand practise various kinds of grafting, whereas budding is widely practised in Malaysia. This is done by budding scions of the selected clones on to rootstock of the mixed and unknown seedlings (Figure 5.3). Stocks are usually one year old and at least 12 mm diameter at 10 cm above ground level. Budwood is cut from the parent tree from branches of similar size to the graft position on the rootstock. Petioles are dropped naturally after cutting off the leaves on the scion tree about a week before budding commences. The most effective budding method is the modified forkert. Forkert budding (or brown budding) was developed in Indonesia, primarily for use on rubber trees. The method involves budding

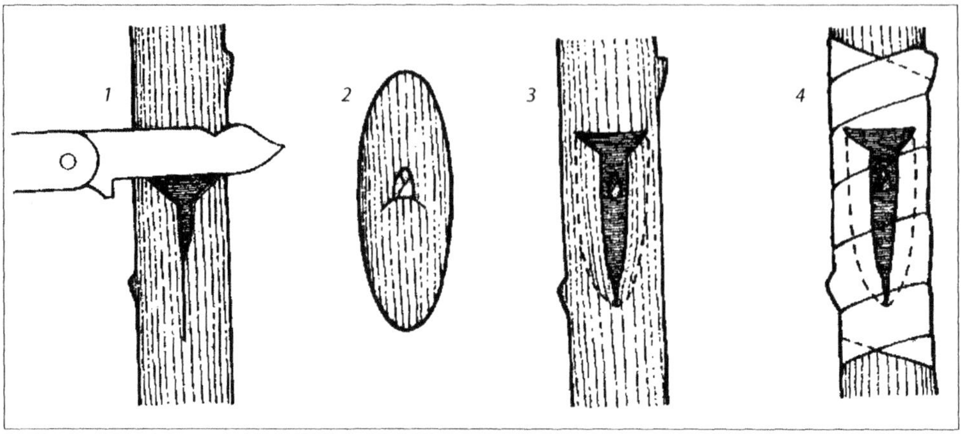

Figure 5.3 *Steps in budding: 1) T cutting; 2) scion bud; 3) inserting the bud into stock plant; 4) wrapped at the budding position*

well-developed, leafless, brown budwood on to actively growing stocks. However 'T' budding appears equally successful.

Patch budding is the method used in large-scale propagation in some countries (Ochse *et al.*, 1961; Reksodihardjo, 1962; Sankhariksha, 1968). The best scions are non-petioled, mature (but not old) wood, and rootstocks are vigorous, one year old seedlings. The buds must not come in contact with water as they will not unite and are likely to rot. The best time to bud is at the onset of the vegetative growth period.

Shield budding has been quite successful in the Philippines (Galang, 1955). Non-petioled, fairly mature budwood is used. The buds are cut to 2.5 to 3.5 cm long. For shield and patch budding, success depends a great deal on the skill of the propagator. In shield budding (Figure 5.4), the stick of buds is held by the upper end and the first well-developed bud nearest this is removed by a shallow slicing cut which begins 1 to 2.5 cm below the bud, passes under the bud and comes out well above it (1). The bud shield should contain a sliver of wood, but should be quite thin and flexible so that it will bend to fit closely over the curved surface of the wood of the stock and be neatly enfolded by the stock rind.

If the knife blade stops just before it reaches the surface at the end of the cut, and the removal of the bud shield is completed by tearing, it will terminate in a long strip of rind that may be used as a convenient handle. This handle is cut off after bud insertion (4). The stock (2) is prepared by making a 'T' incision through the rind down to the wood. The rind is raised and the bud inserted (3). The tail of the handle is cut off exactly at the horizontal incision (4) and the bud is firmly tied with budding tape; polythene strips are normally used (Figure 5.3). The bud must remain firmly tied for two or three weeks, at which time the polythene is removed to prevent the stock becoming constricted.

Top grafting

Cleft grafting on to 9 to 12 month old seedlings is popular in the Philippines (Coronel, 1977). As many leaves as possible should be retained below the graft until there is substantial scion

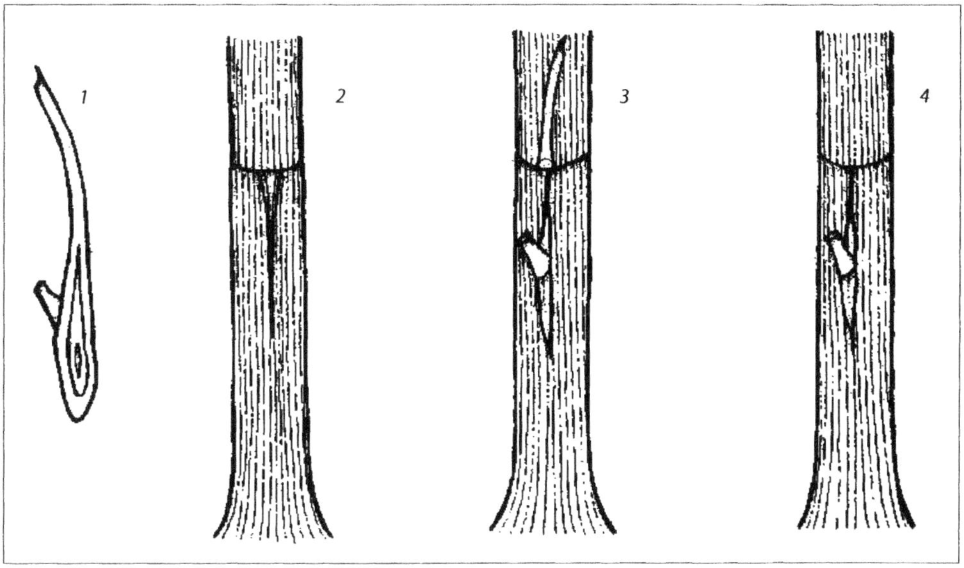

Figure 5.4 *Shield budding: the bud (eye) prepared and inserted into the stock ready for tying*

development. Cleft grafting has been used at Kamerunga Horticultural Research Station with moderate (40%) success (Watson, 1978).

Approach grafting

Approach grafting is common to all durian-producing countries, and is widely practised in Thailand. However, this method is now used to a lesser extent in large commercial nurseries because of the need to construct benches, have stock trees, and have more space and time than required for bud or top grafting. The advantage of approach grafting is that seedlings as young as six months can be used. Thai growers tend to use the cleft approach graft, where the top of the stock plant is removed and inserted as a wedge into the scion branch (Figure 5.5). This method provides a strong tree with a tap root that can withstand wind and water-logged conditions as well as root diseases.

In Thailand, about 90% of durian trees are propagated by a type of bottle graft or approach graft (Figure 5.5), a method developed by local growers. Durian seeds are sown in germinating beds, until the seedlings are about 30 to 60 cm high. They are then lifted and the roots packed with moist coconut coir in a sealed polythene bag. These seedlings are attached to large branches adjacent to the suitable lateral scion branches. The tops of the seedling stocks are removed and inarched to the scion branches in the fashion of a side veneer graft. After approximately 40 to 60 days, or when the new roots from the stock plants are clearly seen reaching the surface of polythene bags, the scions are cut at about the level of the bottom of the seedling's roots. These scion butts, if left over, can develop roots when planted, and the grafted trees have two root systems. However, some growers prefer only one root system, so the scion branches are cut about 5 to 10 mm below the graft union. In Thailand, Chanee cultivar seedlings are commonly used as rootstock for many commercial cultivars.

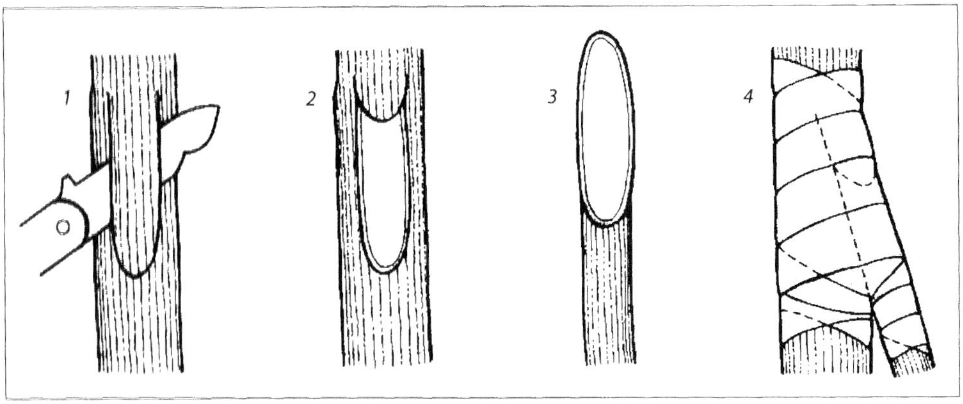

Figure 5.5 *Steps in grafting: 1) cutting stock; 2) stock scion cut; 3) cutting scion; 4) scion inserted into stock and wrapped with polythene*

Top-working

Good varieties of durian are commonly top-worked onto unproductive or poor quality fruit bearing trees in Thailand. The poor quality tree is decapitated at about 1 m in height and is approach grafted with young grafted trees of the desired cultivar on to existing low limbs or regrowth (Boonbongan, 1956).

Alternative rootstock

A relative of the durian, *Cullenia excelsa*, which is native to Sri Lanka, is commonly used as a rootstock in southern India, and has a reputation for imparting greater vigour into scions than durian seedlings (Krishnamurthi and Rao, 1965). *Durio malaccanois* and *D. monsoni* seedlings are now being used in some parts of Thailand as rootstock because they have some resistance to *P. palmivora.*

Cuttings

Stem cuttings have been successfully struck in Malaysia and Indonesia (Watson, 1978). Preliminary observations of the clone D 96 with bottom heat and mist at Kamerunga Horticultural Research Station have not yet proved successful. Plants rooted from cuttings, however, are generally lacking in vigour so this method of propagation cannot be considered commercially viable.

6

Orchard Establishment, Training and Pruning

Orchard Establishment

Site selection

One of the major factors to consider in selecting a planting site is the prevailing climatic pattern in the locality concerned. The range of climatic factors suitable for durian cultivation has been outlined in Chapter 4. The most favourable regions for commercial cultivation are described as being within 12° north or south of the equator, at altitudes of up to 600 or 700 m. The ideal temperature range is from 22°C to 32°C. An annual rainfall of 2 000 to 5 000 mm is desirable, preferably distributed over six to eight months of the year; high humidity is also essential. Exposed sites with excessively strong winds are generally not suitable, unless provision is made for effective windbreaks and the conservation of soil moisture, usually by mulching.

Durian may be grown in marginal areas which do not conform to the above recommended range of conditions. In such areas, extremely good orchard management is necessary, otherwise yields and fruit quality are likely to be affected. In areas where trees are threatened by water stress, irrigation is essential.

Land requirements

Slightly sloping or relatively flat land is preferable, although trees will grow well on land with slopes of up to 10° to 15°. The site must have adequate supplies of water and good drainage. In wetter areas, it is advisable to plant on a slight slope, as durian is very sensitive to high moisture levels in the soil and good drainage is essential for satisfactory growth. On steeper slopes, some form of contour planting or terracing is required to prevent soil erosion, particularly in the early stages of growth. Cover crops may also be necessary as a means of conserving the soil. On a home garden scale with gently sloping land, some form of D-shaped terrace may be built around the base of individual plants to prevent surface erosion and increase the availability of water to the root system.

Soil conditions

These have already been referred to in Chapter 4. Medium clay, sandy loams and alluvial soils with a high organic matter content are preferable. Heavy clay, peat or extremely sandy soils are difficult to manage and will not produce vigorous or high yielding trees. A good soil structure with efficient drainage is essential for satisfactory growth. The site should be assessed for soil suitability by excavating test holes at random over the whole area. These should be at least 1.5 m deep so the soil profile can be examined and checked for the presence of rocky or clay subsoil deposits. Several samples may be thoroughly mixed together in order to obtain a mean sample for analysis.

Durian roots are generally not deeply penetrating, but hard deposits or pans at less than 3 or 4 m deep will seriously impede root development. The soil analysis should assess the main soil elements and should also provide an indication of potential trace element deficiencies. The pH level should also be accurately determined as it may be necessary to apply remedial treatments of dolomite to an alkaline soil, or sulphur to an acid soil.

Other considerations

In areas with lower than optimum rainfall, or where rainfall is irregular, access to an adequate irrigation source must be established before the site is selected. These sources include boreholes, rivers, canals and ponds. The quality of water should be analysed to ensure the absence of salt. Where commercial enterprises are being considered, the costs of pumping water to the site should be included in production costs.

Wherever possible, the site selected should be close to a main road, otherwise considerable expense may be incurred in constructing access roads for transporting produce. Durian is a labour-intensive crop so reliable efficient labour will be essential. Labour costs will also be a significant factor in assessing the feasibility of establishing a commercial orchard.

Orchard Layout

Commercial orchard system

Durian trees require full access to light and any decisions on plant density and tree arrangement must take into account the estimated size of the mature tree, particularly the canopy width and height of the cultivars to be planted. The climatic and soil conditions of the locality should also be considered since both these factors will influence tree vigour. In cooler areas, tree growth will generally be slower than in warmer conditions. In marginally fertile soils, a closer planting distance than normal should be chosen to reflect the resulting reduction in tree size. On sloping land, it will be necessary to adapt the planting distances to take account of the contours.

Planting systems used in commercial orchards in different producing areas vary, reflecting the differences in climate, soil, cultivar behaviour and the management systems used. However, in general, tree rows are planted in a north–south alignment to promote maximum light interception. A square planting system is popular, with trees spaced at either 10 or 12 m apart, giving respective planting densities of 100 or 70 trees per hectare. However, variations on this layout are used in some areas. These include the 'revolving square' system with a central tree established within a square; this tree is normally removed at about 8 to 10 years of age to promote light penetration to the canopies of the remaining trees. Rectangular or diamond pattern plantings, such as are used for lychee, mango and some other fruit crops, are rarely used for durian. Similarly, high density planting systems (in which trees are closely planted for the first 6 to 10 years, often spaced either 5 or 6 m apart and then thinned) have not yet been widely adopted in durian cultivation. The planting systems described above apply essentially to mono-cropped commercial orchard enterprises but similar spacing between trees should be used in home gardens.

Interplanting patterns

In some countries, particularly Thailand, interplanting with rambutan (*Nephelium lappaceum* L.) is common commercial practice. Both trees will crop for several years and when the durian trees overshadow the rambutan, the latter are removed. The traditional practice of planting more than one cultivar is also popular because it results in better fruit setting and therefore a better yield.

Planting systems in major growing areas

The current planting distances and orchard layout patterns followed in the main production areas are as follows:

Thailand

The most popular planting distances are 10 or 12 m apart (100 or 70 trees per hectare) in a square planting layout. A modified quincunx design has also been adopted in some areas. This is essentially a square planting layout (with interior triangles) with plants spaced in a 10 x 10 m or 12 x 12 m square with a central tree that is removed after 8 to 10 years. This arrangement gives a respective planting density of 125 or 88 trees/hectare.

Philippines

The recommended planting distance for durian is 10 to 12 m apart, although distances of 14 to 18 m have also been suggested (Galang, 1955; Ochse *et al.*, 1961; Luardo, 1972). Planting should be carried out at the onset of or during the rainy season.

Indonesia

The planting space is variable. Traditionally, growers have practised a planting distance of 12 to 14 m. Recently, researchers have suggested a closer distance of 8 to 10 m may be more effective. However, taking canopy width into consideration, the recommended spacing for the following cultivars are: Petruk and Sunan at 10 or 12 m apart; Sukun at 6 or 8 m apart; Sitokong at 8 or 10 m apart; Bangkok at 6 m apart.

Malaysia

Spacings of 14.6 to 16.2 m (40 to 47 trees per hectare) are suggested for clonal durian trees in Malaysia (Loh and Lee, 1966). Such wide spacings are recommended in order to prevent canopy competition in mature trees. However, spacings of 8 x 8 m (156 trees per hectare) will produce fruit for five to eight years before canopy competition becomes significant; thinning at that stage would be effective to ensure the most efficient use of land.

Australia

Planting densities in Australia vary between 80 to 120 trees per hectare. A square planting layout is normally used, with a preference for a spacing of 10 m apart for moderately sized cultivars, and 10 to 12 m apart for the more vigorous cultivars.

Land preparation

Thorough preparation of the planting site is essential as the growth rate and ultimate productivity of the trees will largely depend on vigorous early development. If not already carried out in the initial site selection, a soil analysis will be required to determine appropriate treatment for the land both before and after planting.

Site levelling and terracing

On relatively flat land, little levelling will be required, but it may be necessary to use heavy machinery on undulating land to obtain a suitably level surface. This operation will facilitate easier access for spraying machinery and trailers which may be used in the collection of harvested crops.

On sloping land, it will be necessary to construct contour terraces. These are usually retained by earth banks and can be prepared with machinery; on a home garden scale, stone terraces can be constructed.

Earthworks

The levelling operation should be integrated with the construction of access roads and any reservoirs required for irrigation. Both of these operations will involve the use of heavy machinery. The irrigation system to be used should also be planned at this stage and any necessary preliminary installations, such as the laying of main supply lines, carried out. Where the size of the enterprise justifies the use of heavy machinery, the land should be ploughed during the dry season to a depth of about 20 to 25 cm, followed by harrowing. In soils that have been previously cultivated, subsoiling may be required to break up existing compact layers or pans. If the land has not been recently cultivated, it will be necessary to remove all tree stumps, woody shrubs or perennial weeds, and large stones. It may also be necessary to plough and harrow several times to obtain a satisfactory soil condition. The ploughing and harrowing operations should be completed several months before planting. Cover crops can be established once the earthworks are completed.

Planting operations

Planting hole preparation

Planting holes should ideally be excavated up to four weeks before planting begins, normally at the beginning of the rainy season. Holes should be one cubic metre in size in marginally fertile soils but in good quality soils this may be reduced to 60 cm^3. On a commercial scale, wide, tractor-mounted augers can be used for excavating the planting holes, but hand excavation is preferable if labour is available, as it is then possible to separate the topsoil from the subsoil.

To refill the holes, organic material and fertilisers are normally used. The topsoil is usually mixed with well-decomposed organic matter in a ratio of two parts organic matter to one part topsoil and 1 to 2 kg of rock phosphate are also added to the mixture. For relatively infertile soils, 200 to 300 g of a compound fertiliser (NPK 15:15:15) should also be added and the organic matter:topsoil ratio increased to 3:1. Soils that are likely to be subject to magnesium deficiency, as indicated by soil analysis, should also have dolomitic limestone added to the planting hole in a quantity of 200 g per hole.

Planting techniques

If an irrigation system has been installed, planting can begin before the onset of the rainy season but it is usual to plant at the beginning of the rains, preferably during humid cloudy weather. The plants should be carefully transported from the nursery avoiding damage to the root system. Plants are normally transferred in polythene bags, which are removed by slitting them down each side.

Budded and grafted plants must be handled carefully during planting, preferably by holding the base of the stem, as the roots are particularly sensitive. A hole the same size as the root ball is dug in the centre of the hole and the plant is lowered into it. Loose soil (a mixture of topsoil and organic fertiliser or animal manure) is used to fill in the space around the root ball and is firmly pressed with the fingers to ensure close contact of the roots with the soil. Plants which have been correctly handled should recover from transplanting shock within two months.

In Thailand, it is now customary for budded and grafted plants to be raised in a traditional bamboo basket. The main advantage is that there is minimal root disturbance as the plants are planted complete with basket. Root damage on transplanting is one of the main causes of establishment failure.

At the time of planting, it is recommended that the plants be staked to protect them from being damaged by wind or orchard operations. Stakes up to 1.5 m long should be inserted alongside the plant at the time of planting, taking care not to disturb the roots. The young shoots are attached to the stakes by plastic ties. Watering immediately after planting is essential and should be repeated at regular intervals during dry weather, at least three times each week for the first month, to ensure an adequate soil moisture supply for young root systems. About 20 to 40 litres of water per tree should be applied immediately after planting and watering should be continued regularly if the weather is unseasonably dry. A mulch of organic material should also be applied around the base of the plant to conserve soil moisture, particularly if the rainfall is irregular during the planting period.

Shading and wind protection

Durian trees need shade for at least the first 6 to 12 months after planting, particularly in sunny weather. This prevents excessive transpiration from young leaves. A thatch of coconut or oil palm leaves supported on four stakes arranged around the plant is usually adequate but plastic shade netting is frequently used in commercial operations.

Where young plants are exposed to wind, extra protection from excessive transpiration and leaf damage is required. If a rapidly growing windbreak hedge has been previously planted this may be adequate, but as a temporary measure it may also be necessary to protect individual plants from wind. This is achieved by surrounding the young plants with coconut, oil palm or other large leaves. A more effective method is to use 60 to 100 cm high screens of hessian or plastic sheeting attached to pegs and arranged around three sides of the plant as a windbreak. If the supports for overhead shading are sufficiently firm, the temporary windbreak material can be attached to them.

Husbandry practices

During the first month after planting durian should be watered daily, particularly in Thailand where planting usually takes place one month before the rainy season. Mulching with cut grass (or other organic material) is a common practice to conserve moisture. Between 8 and 10 months later, ploughing is recommended to break up the soil around the tree, followed by mulching. After one year, annual ploughing is still necessary but at a greater distance around the dripline. Following ploughing, manure or organic fertiliser should be added.

Different methods of durian husbandry are used according to location, durian variety and the grower's experience. Close observation and everyday care are needed. Some growers believe that chicken manure is best while others recommend a mixture of chicken manure and chemical fertilisers. In addition, one successful grower suggests that watering every four to five days, just to keep the soil moist, is beneficial for young trees. Depending on the local environment, the soil and the age of the trees, both organic and chemical fertilisers can be useful. Under Malaysian conditions, fertiliser is applied quarterly to poor or medium fertility soils, making a total of about 1 kg for a one year old tree. This is increased to 2 kg at two years old and so on. Organic fertilisers are also used on sandy soils depending on the overall performance of the tree. Traditionally, little husbandry apart from clearing the base around each tree during the fruiting season is carried out by growers in mixed orchards in Malaysia and some parts of south Thailand, Indonesia and the Philippines.

Training and Pruning

Growth habits

Durian naturally grows in jungle or semi-jungle conditions, as in the old mixed orchards of South-East Asia. Plantation cultivation consequently presents unusual problems. Most important of these is the difference in growth habits. Trees cultivated under orchard conditions are distinctly pyramidal in shape, whereas under jungle conditions the lower branches are self-pruning and the bearing branches are very high overhead. This results in the development of large weak branches in orchard cultivated trees which often fracture at the trunk causing severe damage. It is clear that if the durian tree is to be grown under orchard or plantation conditions, a great deal of attention needs to be paid to training and pruning techniques.

Training

Asexually propagated durian plants tend to produce several branches close to the ground and this makes training essential during the first few years. Those plants propagated by grafting and inarching (approach grafting) tend to produce suckers from the rootstock. These suckers grow faster than the scion tree and must be removed. Scion branches should be regularly checked for suckers or they may die. In view of the susceptibility of the plants to patch canker (Navaratnam, 1966), no branches should be allowed to develop less than 2 m from the ground (30 cm in juvenile trees). In addition, only three to four well-distributed main branches should be allowed to develop from the trunk. Durian flowers and fruit are produced on large branches in the interior of the tree so keeping the interior open allows better pollination and fruit development. This can be achieved by removing excess lateral branches. All diseased, pest-infested and dead branches should also be pruned regularly.

A previously popular system of training, used particularly in Thailand and Malaysia, was to prune the leading shoot and lateral branches heavily each year during the early stages of growth. This resulted in a very compact low-growing bushy tree. This technique is still practised in some growing areas of both countries, but long, weak lateral branches and poor quality fruit are major disadvantages. Another system, also practised in Thailand, was to cut out all the shoots growing in the centre of the tree. This resulted in the production of long bare branches, reducing the effectiveness of transport of nutrients and promoting early decline of the pruned branches.

The current preference in Thailand is for the modified leader system. The objective is to build up a strong framework of branches by allowing the main shoot to grow to about 70 cm in the first year before the tip is removed. At this stage, three to four strong, lateral, wide-angled branches should be developed. These branches will form the main framework of the tree and are pruned in the second year to encourage the production of secondary lateral branches.

Pruning

Trees left unattended will grow weak branches and may develop epiphytes. These will cause the branches to be prone to pests and diseases, and possibly death. During harvesting, these branches are easily torn or broken, which promotes the likelihood of infestation and disease. Pruning is crucial in reducing these problems. To facilitate early fruiting or to shorten the juvenile period, light pruning only is suggested. Many branches should be allowed to grow naturally, including small branches in the canopy which are normally removed. These branches tend to flower earlier and die off after fruiting. This method is practised successfully in Thailand to produce early fruit. Generally, the Kan Yao durian will flower in the fifth year after planting. However, by using this method the unpruned trees will bear fruit within two years. After this period the yield will decrease, the trees will lose vigour quickly and die sooner. The general objectives of pruning durian trees are as follows:

- To obtain a desirable shape and canopy for convenience of fertilisation, pest and disease control, flowering and harvesting.
- To encourage vigorous growth, resulting in healthy branches that produce good growth and high yields. This can be achieved by removing unwanted and diseased branches and reducing the competition between branches in order to establish framework branches.
- To obtain even distribution of fruit in the tree and to obtain high quality fruit of uniform size.
- To obtain regular flowering and fruiting, that is, to reduce alternate bearing. The tree should be pruned to allow a sustainable number of fruit to develop. Bearing too many fruit will put the tree at risk of losing food reserves and producing few or no flowers or fruit the following season. Flower and fruit thinning, together with the application of suitable fertilisers, will help to maintain food reserves.
- To reduce excessive water loss from transpiration and to maintain water balance within the tree.
- To reduce the chance of disease and insect infestation by removing injured, diseased and unwanted branches. This will increase ventilation and light penetration in the canopy, which discourage infestation.
- To reduce wind damage, as unpruned trees are more susceptible to wind damage than pruned trees with a light canopy.

Pruning juvenile trees

Pruning should be carried out regularly from the time of planting using the following principles:

- Remove any suckers that appear at the base of the rootstock in some cultivars.
- Remove lower branches to allow one main trunk to develop.

- Prune to ensure the branches are not too close together. The branches of a young tree should be 6 to 8 cm apart at the base, increasing to 30 cm apart in the mature tree.
- Three years after planting, top extremely tall trees to enhance lateral growth. However, the lowest branch should remain at least 30 cm above ground level. Topping durian trees to control height is not generally necessary as most trees are pyramidal in shape and only light pruning is needed.
- Remove water sprouts, water shoots and any unhealthy shoots.

Pruning mature and fruit bearing trees

Durian trees bear flowers and fruit on branches, not on terminal shoots like other tropical fruit such as rambutan and mango. Therefore strong healthy branches should be left on the tree. Water sprouts, shoots and suckers that grow into the canopy should be removed. High quality fruit are always found on healthy branches while those fruit borne on the small and unhealthy branches are of inferior quality. Current practice in Thailand consists of three annual prunings:

- The first pruning is done after harvest and involves cutting off the diseased, dried and unused branches. This is to allow for the emergence of new shoots. Branches that have borne too many fruit may become weak and should be removed.
- The second pruning takes place in August or September before the second application of fertiliser. The tree will probably have many water shoots, which should be removed. Fertilisation should be carried out after pruning to stimulate growth.
- The third pruning takes place after fruit setting – this is the last pruning before harvest. Pruning should go ahead when the fruit is about the size of an orange and should target emerging shoots. This pruning is usually done in conjunction with fruit thinning.

7

Nutritional Requirements and Fertilisation

Durian Nutritional Requirements

Durian has nutritional requirements similar to many tropical fruits, although detailed nutritional studies on durian are limited. As a result of this shortage of data, growers in many countries have made use of their own observations and experience in adopting fertiliser regimes for their orchards. Many of these pragmatic recommendations have proved to be very effective.

Small-scale durian growers and home gardeners rarely use chemical fertilisers and most nutritional requirements of their crops appear to be adequately met by mulches and organic manures, particularly chicken manure. Additionally, since many home garden trees are grown from seedlings, their vigorous root systems are able to penetrate extensively into the subsoil, enabling access to an adequate supply of nutrients.

As with many other fruit trees, environmental factors determine to a significant extent the nutritional status of the tree. Good drainage is a particular requirement for the durian, while soil fertility, pH, composition and structure can directly influence fertilisation needs.

Also important are the specific nutritional requirements of some durian cultivars. Some are vigorous in growth and require a higher nutritional input than less vigorous cultivars. Some also have a more extensive, fibrous, lateral-rooting habit than others and therefore require light but relatively frequent dressings of fertiliser to maintain the active growth of sub-surface roots.

There is also a periodicity in fertiliser requirements. Mature plants require higher levels of fertiliser than juvenile ones and it is generally accepted that basal dressings of fertiliser should be applied only after budded or grafted plants are well established in the orchard. Leaf flushes and periods of active growth, particularly in the two months before flowering, should be supported by adequate fertilisation. The period from flowering to fruiting will also make heavy demands on the nutritional resources of the plant. Another critical period for fertilisation is immediately after harvesting and pruning. In well-managed orchards, fertilisation and irrigation during this period will promote new leaf flush which can be of considerable benefit to the following crop (Watson, personal communication). During dry periods when growth is minimal, trees will not normally require heavy dressings of fertiliser.

As with all fruit trees growing in tropical regions, durian orchards are subject to extensive leaching, particularly if there is no cover crop. This will result in the removal of the most soluble nutrients, particularly nitrogen and potassium, and also the loss of minor or trace elements such as zinc, iron and copper. The availability of trace elements is also likely to be influenced by the soil pH level which, if particularly alkaline, may reduce trace elements. Major nutrients such as phosphorus may also be scarce under such conditions.

The level of organic material present in soils is also a factor regulating the availability of soil nutrients. Soils which contain good organic matter will retain significantly more nutrients than those which are lacking. Therefore, in sandy loam soils, it is necessary to compensate for the loss of nutrients by more frequent fertilisation. The use of organic mulches and manures in addition to chemical fertilisers is also beneficial, particularly in well-drained, sandy soils.

The efficiency of orchard management is also important in ensuring trees are adequately supplied with the nutrients they require. In Thailand, it is generally recommended that during the first year, about 100 to 200 g of ammonium sulphate should be applied one month after planting, and repeated before the end of the rainy season. If soil analysis indicates that phosphorus is low, 100 to 200 g of rock phosphate should be applied at planting time. As the trees grow, the quantity of fertiliser applied should be correspondingly increased. When the trees start to bear fruit, it is necessary to apply a compound fertiliser. Nutrient removal during the harvest amounts to 2.4 kg nitrogen, 0.4 kg phosphorus, 4.2 kg potassium, 0.3 kg calcium and 0.5 kg magnesium per tonne of fruit. This indicates that durian trees require large quantities of nitrogen and potassium to sustain fruit production.

A study in Malaysia (Ng and Thamboo, 1967) found that nearly twice as much potassium was removed by the fruit compared with nitrogen. On the basis of production of 6 720 kg of fruit per year (approximately 5 400 fruit), removal of elements was 16.1 kg nitrogen, 2.7 kg phosphorus, 27.9 kg potassium, 2 kg calcium and 3.3 kg of magnesium. Subsequent to these findings, fertilisation has been modified to increase potassium levels in near mature and mature trees. At Kamerunga Horticultural Research Station, the fertilisation programme for durian is split into three side dressings per year to an annual total of 10 g nitrogen, 2 g phosphorus and 17 g potassium per tree per year, up to a maximum of 15 years of age. Additional nitrogen is supplied separately at a rate of 40 g per year of age up to a maximum of 10 years (Watson, 1957).

In Thailand, a general recommendation for fertilising durian trees during the first year of fruiting indicates 1 kg (in total) per tree. That is, 500 g of compound fertiliser may be applied to each tree at the start of the rainy season and an equal amount at the end. The amount of fertiliser must be increased gradually as the trees grow and their yield increases. At the peak of fruit production, an annual application of 2 kg or more per tree of a compound fertiliser may be required to sustain growth and fruit production. Supplementary applications of single fertilisers may have to be applied, containing either nitrogen, such as ammonium sulphate or urea, or potassium, such as muriate of potash.

Animal manures have been and are still used to a great extent in Thailand, Malaysia and Indonesia for incorporation in the planting hole as well as for annual mulches.

Soil and leaf analysis

A comprehensive soil analysis is an essential prerequisite in assessing the suitability of sites for durian cultivation. Based on the findings of this analysis, it is possible to estimate which major nutrients the tree will need so that fertilisation may begin in the year after establishment. The requirements of various soils under different climatic conditions are reflected in existing practices, which are outlined in the last section of this chapter.

After four or five years of growth and once fruiting has begun, leaf analysis may become a routine practice in the durian orchard. Full details of the procedure adopted for collecting lychee leaf samples for analysis have been given by Galán Sauco (1989) and these are equally appropriate for durian. Trees should be inspected at regular intervals and any sign of nutritional deficiency, as evidenced by leaf distortion, necrosis or chlorosis, should be immediately followed by laboratory sampling to identify the cause. Appropriate corrective measures can then be adopted to prevent recurrences.

Nutritional deficiencies

These include major and trace elements and are most likely to occur in alkaline soils, although very acid soils may also be subject to nutritional imbalances. Growing durian on calcareous soils is generally uneconomic because of the lack of trace elements. The following elements are those most frequently associated with durian nutrition:

Potassium deficiency

In the case of potassium deficiency, seedlings and mature trees may both show a form of leaf necrosis (Coronel and Manlisis, 1972). This begins as a brown spot appearing at the tips of mature leaves. As the deficiency increases, the leaf margins become brown and develop a scorched appearance. In severe cases, the necrosis spreads towards the midrib and eventually the entire leaf-blade may become affected and fall off. This disorder appears first on the older leaves, but will eventually affect younger leaves; it usually appears during the dry season, being aggravated by water stress in the plant. It can be prevented by ensuring the potassium status of both the soil and the plant are kept at an optimum level by regular applications of fertilisers containing potassium.

Nitrogen deficiency

Signs of nitrogen deficiency are seen in the stunted growth of branches and leaves. Leaflets are small and chlorotic with slightly curled margins. The root system is also restricted in growth and flower production is reduced. A deficiency may be corrected by applying nitrogenous fertilisers, such as ammonium sulphate and foliar sprays of urea. These are generally effective – especially on young plants.

Phosphorus deficiency

This is rarely serious but when it does occur, leaves may become necrotic and tinged with purple. Root growth may be restricted and flowering affected. The application of triple superphosphate is normally sufficient to correct this deficiency.

Magnesium deficiency

The most visible sign is a reduction in the size of the leaflets which also develop chlorotic areas between the leaf veins. Leaf drop, and poor flower and root development may also occur if the deficiency is acute. Foliar sprays of 1% to 2% of magnesium sulphate and the incorporation of dolomitic limestone into the soil around the base of the tree are normally recommended.

Calcium deficiency

Leaf margins become necrotic and leaf growth is reduced when deficient in calcium. The growing tips of young plants may also be affected, becoming stunted and brown. Limited root development may also occur. Applications of hydrated lime or dolomitic limestone before planting, and to established plants, are normally effective, but care should be taken in applying calcareous materials to young plants – only minimal quantities should be applied. Foliar sprays of calcium sulphate or other calcium compounds are rarely necessary.

Iron deficiency

This appears as an interveinal chlorosis and usually occurs on young plants raised under nursery conditions (Coronel and Manlisis, 1972). The deficiency first appears on very young leaves and is seen as a network of light green veins on a yellow background. Slight deficiencies may be difficult to identify, but in severe cases leaves may become almost completely yellow. They will also be smaller than normal leaves and seedlings may become stunted, eventually dying. Leaves which are seriously chlorotic may contain very low quantities of iron, such as 20 ppm. Affected seedlings can be successfully treated by spraying with a ferrous sulphate solution at a concentration of 500 ppm. Recently, Chelated Sequestrene 138® has been recommended to prevent iron deficiency, particularly in young plants. The solution is prepared by dissolving 10 g of the chemical in four litres of water which is then liberally sprayed around young plants. It has been observed that very alkaline soil conditions can contribute to this disorder, indicating the immobility of the ferrous ions.

Zinc deficiency

This is rarely serious but affected plants may produce small distorted leaves. In Queensland, Australia, four sprays of 2% zinc sulphate per annum have been very beneficial. Although zinc deficiency may not be obvious, it may interfere with the uptake of nitrogen and potassium (Yaacob, personal communication).

Copper deficiency

Leaves become distorted and discoloured but copper-based sprays used to treat various pathogens are normally sufficient to prevent occurrence of this deficiency.

Fertilisation

Types of fertilisers

Granular fertilisers

These may be single fertilisers such as ammonium sulphate – which is widely used in most growing areas – or compound fertilisers containing nitrogen, phosphorus and potassium (NPK fertilisers). NPK fertilisers are formatted differently according to the country they are used in.

Calcium and magnesium are the two elements which are likely to be either completely lacking or restricted in their availability, particularly in alkaline soils. It has become common practice to include one or both of these elements in general NPK formulations. Where deficiencies are serious, dolomite or dolomitic limestone (calcium magnesium sulphate) is added as a separate application. Calcium, in the form of hydrated lime, is also used in some areas. Sulphur, in powder form, can be used on extremely alkaline soil as a means of reducing the pH level. The regular use of ammonium sulphate will also tend to slightly increase soil acidity. Slow-release fertilisers could probably serve a useful purpose in durian nutrition, particularly in those areas subject to high levels of rainfall and subsequent nutrient leaching.

Soluble fertilisers

These fertilisers are mainly used where drip irrigation is available, as the fertiliser is fed into the irrigation system through a regulated feed system. Soluble urea is applied by this method, as are most of the granular fertilisers which have a high level of solubility.

Foliar sprays

Where leaf analysis shows the deficiency of minor elements, this can usually be corrected rapidly by applying foliar sprays of compounds containing the required element. Examples of this are ferrous sulphate, at a concentration of 500 ppm, for correcting iron deficiency, and magnesium and zinc sulphates, at a concentration of 1% to 2%, to correct deficiencies of magnesium and zinc respectively.

Foliar applications of monopotassium phosphate in a 0:52:34 formulation have been recommended for use in Thailand. The application, one to two months before flower initiation, is used to stimulate leaf maturity, reduce leaf flushing and promote flower initiation.

Organic manures

The use of various forms of plant and animal residues as organic manure is widespread throughout major growing areas, especially in home garden production. Relatively little quantitative data are available on the amounts of organic manure applied per tree but there are indications that 30 kg per mature tree is an average level of application in durian orchards in the eastern provinces of Thailand.

Methods of fertilisation

Fertilisation periods vary between producing countries, mainly due to wide differences in soil structure and composition, climatic factors and cultivar variations which respond to these environmental conditions. Since very few investigations into the nutritional requirements of durian have been carried out, most published data relate to specific locations and cultivars. Many sources also quote data on fertilisation which come from the results of practical trials and observations by growers.

During the first two or three years after planting, trees will require a relatively low level of fertilisation, particularly if they have been planted correctly in adequately prepared soil. As the trees develop, these requirements will rapidly increase and it is generally accepted that a five year old tree, on the point of bearing fruit, will require 0.5 kg of NPK fertiliser per year. As growth progresses, this will normally rise to 1 kg per year. It is also common practice to relate the age of the tree to the number of kilograms of NPK required per year, increasing the rate by up to 1 kg per tree per year, applied in three equal applications throughout the year. A ten year old tree would therefore receive applications of 10 kg of fertiliser per year. This empirical practice appears to have proved appropriate in most Thai durian orchards. Critical periods for fertilisation appear to be post-harvest, mid-vegetative growth, before flowering, during flowering and fruit set, and two to four weeks after fruit development. It appears probable that these are the most appropriate physiological periods during which plants require additional stimulus in the form of readily available nutrients.

The three most widely used methods of application are broadcasting, placement and the use of foliar sprays.

Broadcasting

Broadcasting is a commonly used method for applying granular fertilisers. The main restriction is that the fertiliser must only be lightly raked or forked into the top few centimetres of soil. Durian has a fibrous root system that grows close to the soil surface and is particularly sensitive to soil disturbance. It is inadvisable to apply fertilisers close to the trunk of mature trees since the main mass of actively feeding roots is generally found below the fringe of the leaf canopy. A wide band of fertiliser, broadcast below the perimeter of the tree canopy, is the most effective practice. When dolomite or lime must be applied to highly acidic soils to increase the pH, they are usually broadcast over the soil surface and lightly harrowed in. This operation is preferably carried out before planting.

Fertiliser placement

In some regions, particularly southern and eastern Thailand, organic manures are applied in a trench placement technique. A trench about 30 cm deep and 50 cm wide is excavated below the edge of the tree canopy. This trench is filled with well-decomposed manure which is then covered with the excavated topsoil (Yaacob, personal communication). Also in Thailand, granular fertilisers are often applied either during the dry season or the fruiting period by placing them in a furrow 5 to 10 cm deep which is excavated around the tree below the canopy edge. The fertiliser is mixed with the excavated soil which is then replaced in the furrow.

Foliar application

Deficiencies of iron, magnesium and other elements occasionally occur, particularly in highly alkaline soils. The use of foliar sprays, such as ferrous sulphate at a concentration of 500 ppm, is the most frequently effective treatment. Signs of deficiency and the use of foliar sprays have already been referred to earlier in this chapter. These sprays are usually applied to young trees using a knapsack sprayer. For more mature trees, tractor-mounted sprayers with a high nozzle output or motorised knapsack sprayers with long lances are required.

Recommendations for specific countries

Although very little data on the effects of various fertilisers on the growth and development of durian have been published, the following recommendations are those normally followed in the countries in which durian is most widely grown:

Malaysia

The recommended applications for marginal soils include an NPK mixture in a ratio of 14:4:4 up to the first harvest, followed by subsequent applications of a 12:4:7 mixture. Application rates begin with dressings of 0.1 kg per tree per annum at planting, increasing progressively to 4 kg per annum when the trees are 20 years old (Kanapathy, 1976). Yaacob (personal communication) states that the current Malaysian commercial practice is to apply 15:15:16 or 16:16:16 NPK fertilisers during the vegetative growth stage, and to follow this during the fruiting period with dressings of NPK at 12:12:17 - 2 (calcium) and 13:13:21. The applications begin at a rate of about 0.5 kg per tree per annum for two to three year old trees, split into two or three dressings, and the rates are increased by 0.5 kg per tree per annum as the trees mature (Pongsrihadulchai, 1984). Yaacob (1983) states that once established (after about the seventh year), durian trees do not need heavy fertilisation. Application of small amounts of fertiliser after each harvest for maintenance and counteracting the amount lost through leaching and surface wash may be all that is required.

Philippines

A dressing of 1 kg of ammonium sulphate plus 1 kg of an NPK mixture at a concentration of 12:24:12 per tree, applied post-harvest, is a commonly used treatment. An additional 1 kg of ammonium sulphate per tree, applied at the end of the rainy season, is also a popular practice.

Coronel (1983) gives general recommendations for durian fertilisation. During the first year, about 100 to 200 g ammonium sulphate may be applied one month after planting and an equal amount before the end of the rainy season. As the trees grow, the quantity of fertiliser applied should be correspondingly increased. When the trees start to bear fruit, it is necessary to apply a compound fertiliser.

At the peak of fruit production, an annual application of 2 kg or more per tree of a compound fertiliser may be required to sustain growth and fruit production. Supplementary applications of single fertilisers containing nitrogen (eg ammonium sulphate or urea) and potassium (eg muriate of potash) may have to be made.

Thailand

In eastern and southern Thailand, where soils are very sandy and well-drained, organic manures are widely used. They may be applied either at the beginning of the rainy season or after harvesting and pruning. The usual method of application is trench placement. Recommended fertilisers at various growth stages are as follows :

First two years after planting

During this juvenile phase, a compound NPK fertiliser at a concentration of 15:15:15 should be applied, 300 to 500 g per tree. Three or four split applications (or at least two) of 100 to 200 g per tree are recommended. Trench placement is generally used for these applications. Organic manure, 5 kg per tree, should also be applied.

During vegetative growth

After harvesting, durian trees shed their leaves. Applying fertiliser at this stage will strengthen the trees and prepare them for the next flowering season. Thus, compound fertilisers, such as 15:15:15 NPK, at 3 to 6 kg per tree are used, with a supplementary application of 15:15:15 NPK three to four months later.

Before flowering

At this stage the fertilisers applied should contain less nitrogen while quantities of phosphorus and potassium are higher, eg 9:24:24 NPK at 3 to 6 kg per tree in August or September. This assists in flower development. Foliar application of a high phosphate fertiliser, eg 10:52:17 NPK, may assist in hastening flowering.

At fruit development

After fruit setting, the trees need large quantities of nutrients to support the growing fruit. A lack of nutrients may result in heavy fruit drops or stunted fruit. Sufficient water and fertilisers are essential to the trees at this stage. Fertilisers high in potassium are recommended. For example, 13:13:21 or 14:14:21 NPK. This should be applied in three applications at two to four week intervals at 1 to 2 kg per tree per application. Potassium is important in order to increase the quality of harvested fruit.

Indonesia

Recommended fertilisation is as follows. Three months after planting, NPK (15:15:15) at 200 g per tree is applied as a supplementary fertiliser. After this, supplementary fertilisation with NPK should be done routinely every four months until the tree is three years old. Organic compost or manure of 60 to 100 kg per tree may be applied in the dry season. This is done by digging a ditch around the trunk right under the canopy edge. In the fourth year, the dosage of NPK should be raised by 20% to 25% from 500 to 600 to 625 g per tree. By the flowering period, durian needs NPK in proportions of 10:30:10. Applications should take place two to four weeks before flowering.

8

Irrigation Requirements and Techniques

Irrigation Requirements

In some durian production areas, irrigation is not widely used as the rainfall distribution is considered adequate for the needs of the plant. This situation, however, applies mainly to areas where production is based in home gardens or small orchards. In countries such as Malaysia, Thailand and Indonesia, some form of irrigation is necessary if optimum yields and fruit quality are to be achieved. Irrigation is especially essential in the production of export fruit as trees in these countries are more likely to be subject to soil moisture stress, particularly during the fruit development phase. This may affect the marketability of the fruit.

Factors affecting irrigation requirements

Vegetative growth period

Throughout this active growth period, trees require abundant supplies of ground water in order to carry out normal metabolic processes. In very sunny weather, particularly when humidity is low, levels of evapotranspiration may be very high. Such conditions may be harmful, particularly to young trees and may result in a reduction of yield. However, this situation may be partially prevented by the use of mulches and cover crops.

In most areas of production, the irrigation regime has been devised as a result of practical experience and observation. In parts of Thailand, particularly in the east, open pan evaporation readings are used to assess water requirements, although many growers tend to slightly exceed recommended irrigation levels (Ruamsuke, personal communication). As a general rule, trees are irrigated every two to three days during periods of drought except during the flower initiation period.

The levels of irrigation applied to actively growing trees should be related to the size and age of trees, as well as to the soil type and climate. Mature trees with a high leaf surface index will obviously require more water than recently transplanted trees. There is also an apparent variation in the amounts of water required by different cultivars, but although no data are available on this subject, it can be assumed that the more vigorous types of tree will transpire more than compact bushy trees.

During periods of severe drought, leaf margin necrosis and wilting may occur. If up to 25% of leaves show signs of yellowing and incipient leaf drop, trees should be irrigated immediately. If the leaves do fall, the affected trees will be unable to produce flowers (Ruamsuke, personal communication).

Flower initiation period

During the period leading up to flower initiation, which normally coincides with the end of the rainy season, irrigation is not usually required as it is essential for a certain amount of water stress to occur in order to terminate vegetative growth while maintaining floral development. This involves an alteration in the internal metabolism of the plant including the hormone balance and carbohydrate/nitrogen ratio.

If it should rain before flowers are initiated, it is possible that leaves instead of flower panicles may be produced on some branches. In this situation, the best procedure is to apply fertiliser and continue irrigating to encourage leaf flush. Flowers may then form later in the season when the leaves have matured. No research appears to have been done on the possibility of inducing out-of-season crops, especially in lower rainfall areas, by adjusting irrigation schedules. In Thailand, some growers have learned from their own experience to vary flowering and fruiting periods with some success.

Flower development period

Trees require moderate levels of irrigation once the flowers have been initiated. It has been suggested that the level of irrigation should be reduced at anthesis to increase the fruit set percentage (Hiranpradit *et al.*, 1998). High levels of rainfall during flowering have been observed to result in poor fruit set. This is possibly because pollinator activity has been reduced, or because flower parts have been damaged by rain.

Fruit development period

Adequate soil moisture is essential during fruit growth. Insufficient rain can cause fruit drop. However, if too much water is applied, young shoots will emerge and the competition that results may also cause fruit drop. As the fruit approach maturity, the amounts of water applied to the tree should be reduced; this will increase the pulp quality, giving better colour and less moisture.

Soil and water factors

Very little research has been carried out on the root development of durian trees, and the extent of root growth under various soil conditions can only be estimated. Seedlings seem to have a much more extensive root system than clonal cultivars and it can be assumed that vigorous cultivars have more strongly developed roots than more compact ones. Trees with a low water-retaining capacity, grown in sandy soil, will obviously require more regular irrigation than those established in heavier soil, but no data are available on the relative frequency or level of irrigation required for trees grown on various types of soil. It is generally agreed that durian roots are sensitive to excessively high ground water levels and that adequate drainage is essential for satisfactory growth. Salinity is known to be harmful to the trees and areas in which only brackish water is available for irrigation should be avoided.

Irrigation volume

Durian trees require a certain amount of water for growth and development, for physiological activities, and to replace the water lost during growth. In general, the two major routes in which water is lost from the tree are: a) transpiration of water from the surface of the leaves; and b) evaporation from the soil around the trees or from the surface of the tree during and after watering. Therefore, the amount of water required by a tree varies according to the stage of growth and environmental changes in the growing area, such as soil moisture, the orchard microclimate and orchard management. In order to determine the water requirements of a durian tree, Hiranpradit *et al.* (1998) employs the formula:

WR = Kc x ETp where:

WR = water requirements
Kc = crop coefficient which varies according to crop species and age
ETp = potential evapotranspiration (mm/day)

ETp can be calculated by the Modified Penman Method which is derived from climatic data. Table 8.1 shows potential evapotranspiration data over one year in the eastern provinces of Thailand, and Table 8.2 shows the crop coefficient (Kc) during the different stages of growth.

Once the ETp and Kc of the crop during the different stages of development have been determined, we can calculate how much water the crop will require as follows :

Step 1 For a durian tree in Chanthaburi during later fruit development in March, water requirements are established using the following formula:

WR = Kc x ETp

Kc of durian during later fruit development = 0.85 (Table 8.2)
ETp in March in Chanthaburi = 4.5 mm/day (Table 8.1)

WR = Kc x ETp
∴ WR = 0.85 x 4.5
= 3.83 mm/day

Table 8.1: Potential evapotranspiration (ETp) over one year in the eastern provinces of Thailand (Hiranpradit et al., 1998)

	Month (mm/day)											
Province	*Jan*	*Feb*	*Mar*	*Apr*	*May*	*Jun*	*Jul*	*Aug*	*Sep*	*Oct*	*Nov*	*Dec*
Nakhon Nayok	*4.1*	*5.3*	*5.3*	*5.4*	*4.9*	*4.7*	*4.4*	*4.7*	*4.3*	*4.1*	*4.3*	*5.7*
Prachin Buri	*4.3*	*5.3*	*5.2*	*5.4*	*4.9*	*4.5*	*4.3*	*5.1*	*4.2*	*4.2*	*4.5*	*4.1*
Chon Buri	*4.2*	*5.3*	*5.4*	*5.7*	*4.9*	*5.0*	*4.6*	*4.4*	*4.4*	*4.2*	*4.4*	*4.2*
Rayong	*4.2*	*5.2*	*4.9*	*5.3*	*4.6*	*4.7*	*4.5*	*4.2*	*4.3*	*4.1*	*4.3*	*4.2*
Chanthaburi	*4.1*	*4.8*	*4.5*	*4.9*	*4.3*	*4.1*	*3.9*	*3.7*	*3.9*	*4.0*	*4.3*	*4.1*
Trat	*4.0*	*4.6*	*4.4*	*4.6*	*4.2*	*4.0*	*3.8*	*3.6*	*3.9*	*3.9*	*4.1*	*4.0*

Table 8.2: Crop coefficient (Kc) during the different stages of development of different fruit trees (Hiranpradit et al., 1998)

	Kinds of fruit trees				
Stages of development	*Durian*	*Rambutan*	*Mangosteen*	*Citrus*	*Others*
Vegetative growth	*0.60*	*0.60*	*0.60*	*0.65*	*0.75*
Floral initiation	*0.00*	*0.00/0.60**	*0.00*	*0.00/0.65**	*0.00*
Floral development	*0.75*	*0.75*	*0.75*	*0.70*	*0.75*
Fruit setting	*0.50*	*0.75*	*0.75*	*0.70*	*0.75*
Early fruit development	*0.60*	*0.80*	*0.80*	*0.70*	*0.75*
Later fruit development	*0.85*	*0.85*	*0.85*	*0.75*	*0.80*
Fruit maturity	*0.75*	*0.85*	*0.85*	*0.75*	*0.75*

** The floral initiation of rambutan and citrus requires a dry spell before watering to stimulate the process. However, these calculations can be used to indicate the amount of water needed for irrigation. Growers should adjust this amount according to crop response.*

Therefore the water rate of a durian tree during the later fruit development stage in Chanthaburi is 3.83 mm per day. From this we must determine how much water, in total, is necessary according to the tree's canopy as follows :

Step 2 If the area of the tree canopy is 10 m^2, then the tree needs to be irrigated at a rate of

Daily requirement = WR x a

Where WR is the water rate and a is the area of the canopy in square metres.

i.e. Daily requirement = 3.83 x 10

= 38.3 litres/tree

These calculations may not be exact in every production season as the ETp value given in Table 8.1 represents the average over only the last 25 years, as recorded by the Meteorological Department of Thailand.

Irrigation frequency

Once the tree's irrigation requirements have been established, the next step is to consider how frequently it should be watered. This will depend on the following factors:

Plant factors

1.1 Those plants that require frequent watering are as follows:
- shallow root systems with minimal root distribution
- susceptible to drought
- major growth periods are during the dry season
- trees are harvested for leaves, flowers or fresh fruit

1.2 Those plants that do not require frequent watering are as follows:
- deep root systems, densely spread roots
- resistant to drought
- major growth periods are during the rainy season, or while the soil is very moist
- trees are harvested for seeds or dry fruit (ie nuts)

Soil factors

2.1 Those soils that require frequent watering include:
- shallow top soil
- low penetration of water through soil surface
- coarse soils
- soils that are not well-distributed
- soils that have too deep a water table for the roots to use
- soils with a low moisture content

2.2 Those soils that do not require frequent watering include:
- deep top soil
- rapid penetration of water through the soil surface
- fine soils
- well-distributed soils
- soils with a water table at a level that the roots can reach
- soils with a high moisture content

Climatic factors

3.1 The climatic conditions that require frequent watering include:
- high temperatures
- low humidity
- strong winds
- clear sky, bright sunshine
- very little or no rainfall

3.2 The climatic conditions that do not require frequent watering include:
- low temperatures
- high humidity
- cloudy conditions
- rainfall of 1 500 to 3 000 mm per year

Orchard management factors

4.1 Orchard management resulting in a need for frequent watering includes:
- allowing too much fruit on the trees, ie little or no fruit thinning
- application of fertiliser and water to the surface of the soil
- shallow planting or raised-bed planting
- no mulching
- lack of water-absorbing materials to the soil

4.2 Orchard management which precludes frequent watering includes:
- adequate fruit thinning, ie reasonable number of fruit on the tree
- application of fertiliser and water below the surface of the soil
- deep planting
- mulching
- use of water-absorbing materials in the soil

Table 8.3: Effective root depths for water and nutrient absorption for some tropical fruits (Hiranpradit et al., 1998)

Fruit species	*Effective root depth (cm)*
Durian	*20–30*
Rambutan	*30–60*
Mangosteen	*90–120*
Citrus	*120–150*
Other tropical fruits	*100–200*

The above information suggests that frequent watering in small amounts will be more beneficial to durian trees than applying large infrequent quantities of water (Figure 8.1).

Irrigation Techniques

Based on observations of their effectiveness, several irrigation systems have been introduced and adopted in various durian production areas in the eastern provinces of Thailand, such as Rayong, Chanthaburi and Prachin Buri. However, little research has been carried out to determine seasonal water requirements, frequency and optimum levels of irrigation. A traditional method of irrigation is flood irrigation, which consists of saturating the soil around the trees, usually with water from a flexible hose. This is now considered to be wasteful and is less widely used.

Basin irrigation

This fairly traditional method of water application is still used in many production areas. A shallow trench is excavated around the base of individual trees, a little way out from the trunk, and water is pumped into this trench through plastic delivery pipes. Alternatively, water can be supplied through furrows from a main channel or pipe. Considerable experience is required to operate this system to ensure adequate irrigation. Factors which should be

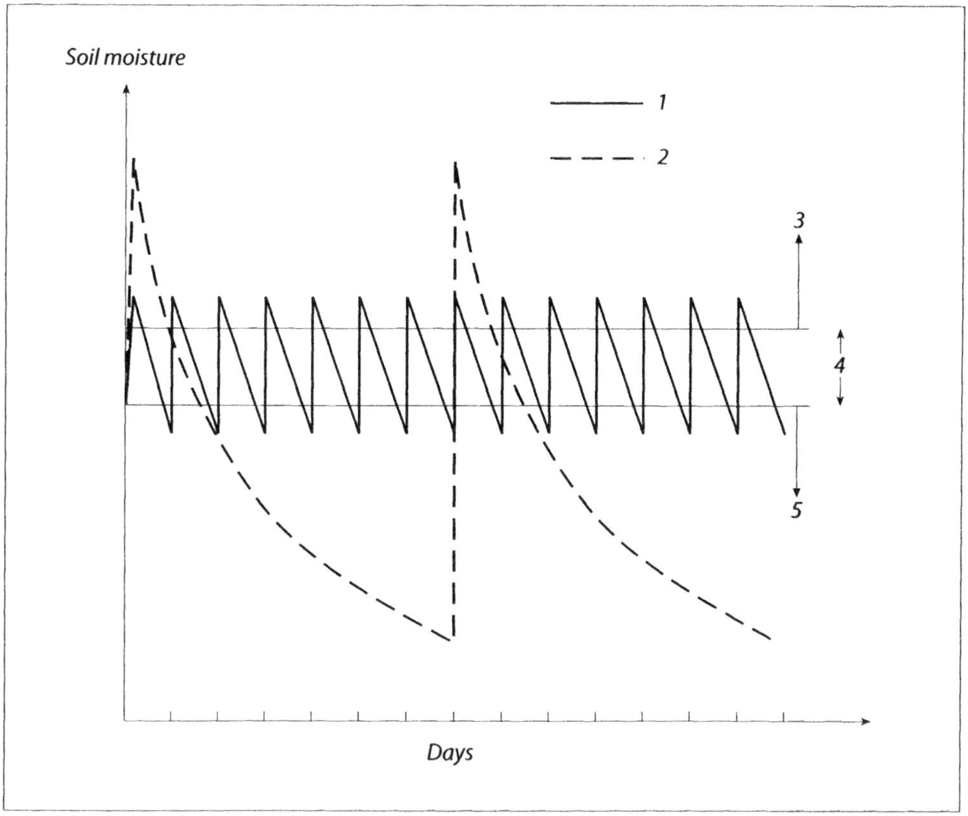

Figure 8.1 *Relationship between the frequency of irrigation and soil moisture*
1. watering everyday
2. watering at seven day intervals
3. an excess of moisture causing oxygen deficiency
4. adequate soil moisture for good growth
5. inadequate soil moisture

taken into consideration include the soil type involved, and the vigour and age of the tree. One disadvantage is the possibility of salt accumulation as water evaporates from the basin surface.

Sprinkler irrigation

This system is widely used in some countries, particularly Thailand, but mainly in areas with a plentiful water supply. The water from rivers, ponds or boreholes is pumped under pressure through main supply pipes to sprinklers, usually through fixed lateral pipes. The number of sprinklers used will depend on the orchard layout and tree density; generally two sprinklers per tree will be required to ensure thorough irrigation of the root zone.

Drip irrigation

This method is appropriate for use in areas where water supplies are limited. The water is usually supplied from an overhead tank to which filters are fitted. The nozzles are spaced at intervals along a plastic pipe which is aligned along the row of trees. Two or more nozzles are required for each tree and these should be checked regularly for blockages. As the trees mature, it is customary to increase the number of nozzles per tree, particularly in areas where orchards are established on sandy soils.

Mini-sprinklers

Mini-sprinklers, or microjet sprinklers, are currently popular in many production areas. In Thailand, various types are in use; some have been imported from countries such as Australia and Israel where they are widely used, but some are locally manufactured.

Current irrigation practices in Thailand

The manipulation of crops by irrigation is widely practised in Thailand. This is especially the case when early flowering is induced by withholding water before the initiation stage. The major durian harvesting season is from May to June, and crops that ripen outside this period are likely to command premium prices. In particular, crops that ripen earlier than normal seem to gain the best price advantage. Those trees growing on sandy loam soils have been observed to respond most favourably to this method of forced flowering. Some cultivars, such as Kradum Thong and Chanee, are more responsive to the withholding of irrigation to induce flower initiation than others, such as Kan Yao or Kob.

At the end of the rainy season, some growers irrigate to hasten the development of young leaves. Irrigation is then stopped for one month or more and the growers clear the ground around the base of the trees, removing all mulch and weed growth to dry out the soil. How long irrigation is withheld depends on soil type, soil condition and fertility, and also on the cultivar and age of the trees. Irrigation is then resumed when flowering begins.

9

Other Production Practices

Flower and Fruit Thinning

In general, durian will produce two to three flushes of flowers a year (Plate 9.1), but the thinning of the flowers is usually done either in the first and third flushes, leaving the second unthinned, or in the second flush, leaving the first and third unthinned. This varies, depending on what time the grower wishes to harvest the fruit. If all the flowers are left unthinned, however, intense competition will cause many of the flowers in the first flush to drop.

Restricting the number of fruit on the tree is the most important way to ensure the production of good quality durian. Fruit thinning at different stages of development will help to maintain the health of the trees. It will also help to reduce the need for agricultural chemicals to get rid of pests and diseases, as well as making it easier to work in the orchard. The number of fruit left on the tree after thinning depends on the size of the bearing branches. In general, on the larger branches in the lower part of the tree, fruit should be left towards the end of the branch, whereas on the smaller branches in the upper parts of the tree, the fruit nearer the trunk should be left. On larger trees, fruit are left on the secondary and tertiary branches, and not on the primary branches. On medium-sized and small trees, some fruit are allowed on the primary branches, as well as on the secondary branches. It is recommended that healthy trees over ten years old should bear 50 to 150 fruit at one time.

Fruit thinning should be carried out twice per crop. The first thinning should be done 30 to 35 days after full bloom or when the fruit reach the size of a lime. Fruit that are abnormally shaped, injured, pest-infested or disease-ridden should be thinned, as should fruit that are too close together in a cluster. The degree of thinning depends on the variety, age and vigour of the trees, but in general no more than two fruit per cluster should be left on the tree.

A good example of the influence of thinning on fruit size was observed in the Chanee cultivar growing in the Rayong province, Thailand. During flowering, between 50 000 and 60 000 flowers were produced by a mature (14 to 15 year old) tree. For an unthinned mature flowering tree, the distribution of fruit according to size was as follows: 20% large fruit, 40% medium-sized fruit, and 20% small fruit. The average weight of these fruit was 1.5 kg. By comparison, similar aged trees of the same variety, which had been thinned until only 120 fruit remained on the tree, produced fruit with an average weight of 2.7 kg. If only 60 to 70 fruit were retained on the tree, the average fruit weight was greater than 5 kg but these were of poor quality. For the Kan Yao variety, 60 fruit should be left on the tree.

Flower Forcing

Durian is generally a seasonal crop and at peak harvest time the price of the fruit is low. Extending the harvesting period and forcing the tree to produce an early crop are the dreams of many durian growers who want their fruit to fetch high market prices. The methods used to produce unseasonable fruit are not easy to implement. A good understanding of the growth cycle and growth habits of the durian tree, together with the skilful use of husbandry techniques in the orchard, are essential for achieving these goals.

The use of paclobutrazol

The durian harvesting period is rather short in Thailand; it lasts from April to May in the central and eastern provinces, and from May to June in the south. Thus, the plentiful supply of durian lasts only four months in Thailand, whereas in Indonesia and Malaysia the harvesting period lasts five to six months. The techniques used for producing off-season durian are eagerly sought after by many growers in Thailand (Subhadrabandhu and Tongumpai, 1990). Thai scientists are now investigating various methods of inducing flowering in durian, often using plant growth regulators like paclobutrazol (Subhadrabandhu and Kaiviparkbunyay, 1998) to retard vegetative growth and enhance flowering. However, the effectiveness of this chemical in inducing flowering is inconsistent, since flowering is a very complex phenomenon which involves many factors.

Hasan Mad and Karim (1990) studied the effect of paclobutrazol on the flowering and fruiting of durian in Malaysia. They applied three different concentrations (0 g, 5 g and 10 g of active ingredient per two litre solution) to each tree as a collar drench at different times of the year. The results showed that paclobutrazol appeared to hasten the onset of flowering but the effects seemed to vary depending on the date on which the chemical was applied. However, the intervals between the appearance of the first flower cluster, the first fruit drop, and the last fruit drop appeared to be unaffected (Table 9.1).

Paclobutrazol has a positive effect on the number of flowers produced. The untreated trees produced very few flowers, while those treated with paclobutrazol flowered profusely (Table 9.2). The results indicated that there was an optimal concentration at which paclobutrazol should be applied. Trees which were treated with 10 g active ingredient had

Table 9.1: Application dates of paclobutrazol, and its effect on the flowering and fruiting of durian clone D 8 (Hasan Mad and Karim, 1990)

	Date of application (1987)		
Fruiting stage	*23 July*	*20 August*	*24 October*
1. First flower appeared	4 January 1988	10 January 1988	14 February 1988
2. Interval between application of chemical and appearance of first flower cluster (days)	127	140	93
3. Date of first fruit drop	7 May 1988	15 May 1988	14 June 1988
4. Time interval between first flower and first fruit drop (days)	123	124	120
5. Date of last fruit drop	12 June 1988	8 June 1988	18 June 1988
6. Interval between first fruit drop and last fruit drop (days)	35	23	35

Table 9.2: Effects of paclobutrazol on flowering and fruiting of durian (Hasan Mad and Karim, 1990)

Paclobutrazol concentration (g a.i./2 l water)	Plant response type	Date of application (1987)		
		23 July	*20 August*	*24 October*
0 g a.i.	No. of flowers produced	261	0	0
	No. of fruit produced	10	0	0
5 g a.i.	No. of flowers produced	1895	457	1285
	No. of fruit produced	20	12	23
10 g a.i.	No. of flowers produced	524	977	918
	No. of fruit produced	13	30	15

fewer flowers than those treated with 5 g active ingredient, suggesting that under the Malaysian climatic conditions the optimal level of paclobutrazol is less than 10 g active ingredient per tree (Hassan Mad and Karim, 1990).

The number of fruit left on the trees showed that only a small percentage of the flowers fully developed into fruit. Treated trees bore significantly more fruit than untreated ones but the concentrations of paclobutrazol applied (5 vs 10 g a.i.) had no noticeable effect on the number of fruit obtained (Table 9.2).

As in the case of the number of flowers, the number of fruit which developed as a result of the application of paclobutrazol was not affected by the date of application (Table 9.3).

The study in Malaysia and Thailand using paclobutrazol with durian indicates that this chemical has a positive effect in promoting flowering and fruiting. Although the role of this chemical in these two processes is not yet fully understood, the data obtained indicate as before that the effect of the chemical is mainly related to the process of flower formation, rather than the subsequent process of development.

Table 9.3: Effect of application date and concentration of paclobutrazol on the number of flowers and fruit produced by durian (mean square value) (Hasan Mad and Karim, 1990)

	Degrees of freedom (df)	No. of flowers	No. of fruit
No. of replicates	3	2 042 301	320
Date of application	22	527 041	12
Error (a)	6	311 233	119
Concentration of chemical	2	3 892 803	933*
Concentration X date	4	945 013	291
Error (b)	18	1 527 263	283
Total:	55		

*Significant at 5% level

From a practical point of view, paclobutrazol can be used in the short term to induce fruit production outside the normal fruiting season, ie off-season production. Before this chemical is recommended for widespread use, however, it is essential that further studies be carried out on its physiological role in order to fully understand its effects on the growth and development of the treated trees. More studies on its residual effects on tree performance in subsequent years must be carried out before its impact may be fully understood.

Other techniques in flower forcing

Durian trees that are ready to flower and are subjected to more than ten days of dry conditions will produce many flower clusters at one time. This makes the management of fruit setting much easier. However, if the trees are not ready to flower while the climatic conditions are suitable, or if the trees are ready to flower but the climatic conditions are unsuitable, the trees will produce fewer flower clusters in many flushes. This is particularly evident on the larger trees with big branches. Proper management is necessary in order to force the trees to produce many flowers at one time, or at least with the lowest number of flower flushes.

The trees are not ready to flower but the climatic conditions are conducive to flowering

This situation occurs in prolonged dry periods, ie when there is no rain for more than ten days but the trees are not yet ready to flower. The leaves are not yet fully mature and the terminal shoots are not standing erect, ready for flowering. Under these conditions, flowering can be hastened by using a mixture of this formula: 600 g glucose solution + 20 cc humic acid + NPK fertiliser at 15:30:15 + fungicide and surfactant mixed in 20 litres of water, sprayed together with mepliquat chloride (poka D 50 cc/20 litres of water). This will promote the maturing of leaves and terminal shoots, as it reduces the level of gibberellin in the tree to a certain extent. In the case of extended dry conditions, irrigation to moisten the soil surface every day will help to induce a high density of flowering at the same time.

The trees are ready to flower but the climatic conditions are not conducive

This situation occurs when dry conditions persist for three to four days and are interrupted by rain, which can lead to the development of vegetative growth. This will result in the formation of few flowers in many flushes. Careful orchard management in order to increase water stress for the trees and stimulate flower bud development will reduce this problem. The application of 150 g potassium nitrate, together with 40 cc Florigen® mixed in 20 litres of water sprayed to cover the tree canopy, is recommended to induce flowering density and reduce the number of flushes (Hiranpradit *et al.*, 1998)

Success in producing early season durian has been achieved by some leading durian growers in the Chantaburi province in eastern Thailand. The general techniques can be described as follows:

Pruning

This must be carried out 15 to 20 days after harvesting. This is done in order to reduce the number of vegetative shoots and remove diseased ones.

Fertilisation

Fertiliser (NPK at a ratio of 9:23:24) is applied at a rate of 2 to 5 kg per tree. The amount of fertiliser applied depends on the age of the tree.

Foliar-applied nutrients and growth regulators

Nutrients and growth regulators are applied to the foliage to induce new shoots to grow, so that they can help manufacture food at this early stage. A foliar spray of NPK (20:20:20 or 18:18:18), together with gibberellin solution (1 tube gibberellin in 10 litres of water), is used to obtain healthy shoots. Once the new shoots are established, a mixture of fertilisers high in phosphorus, potassium and minor elements is sprayed every two or three weeks to hasten the development of vegetative shoots.

Fruit Setting

Fruit setting is a very important factor in determining yields. Good orchard management techniques that aid fruit setting include:

Flower thinning

This is carried out to ensure the flowers are of the same age. Young flowers that were flushed in small numbers are thinned out, leaving flowers from the same flush on the tree. If too many flowers form, they must be thinned out to about 20 flowers per cluster. Thinning should be performed when the young flowers have been out for about 30 days. The remaining flowers should be properly spaced on the branch.

Fertilisation during flower development

The nutrients known to aid fruit setting by stimulating growth and development of the pollen tubes are potassium, magnesium, calcium and boron.

Irrigation management to increase fruit set and development

Watering should begin when the flowers have been out for about 20 days and should continue until the petals turn white. During this period, a rather high amount of water (Kc=0.85) should be applied to the tree. From the white petal stage on, the amount should be lowered by about 40% (ie at Kc=0.50) but applied more frequently to maintain soil moisture from that time until one week after full bloom. When the tip of the stigma that is attached to the young fruit has dried up and turned brown (ie one to three weeks after full bloom), the amount of water should be increased slightly (ie Kc=0.6). The amount of water applied should be increased again up to Kc=0.7 three weeks after blooming, or when the tip of the stigma has dried and turned black. This level of soil moisture should be maintained until the fruit is five weeks old. This type of water management will reduce flower and young fruit drop which can be serious for durian growers.

When moderate to heavy precipitation occurs during flowering, water management as described above cannot be carried out. It is then suggested that growers try to maintain soil moisture at a uniform level, especially beneath the tree canopy. This can be done by applying water when the soil starts to dry out, while removing plant debris and fallen flowers from the soil surface beneath the canopy to increase soil surface ventilation. This practice can help minimise the dropping of flowers and young fruit to a certain degree, though it is less effective than the water management techniques mentioned earlier.

Pest and disease control

The main insects that damage durian flowers include thrips, mites, aphids and caterpillars (which eat flower pedicels). Diseases include *Phytophthora palmivora* and anthracnose (which cause flowers to rot and dry out, respectively). Care should be taken to control these pests and diseases during flower development.

Pollination

Under natural conditions, insects are the means by which durian flowers are pollinated. With the decrease in the numbers of these pollinating insects, however, natural pollination is becoming less successful. This affects fruit setting, especially in the case of the Chanee cultivar, the fruit of which is successfully set using cross-pollination. Artificial pollination is becoming a common practice among some growers as the fruit created through cross-pollination develop a good size and shape, with a high quality aril and a good taste that does not differ from that of fruit developed by natural pollination. Although these fruit may have more seeds, the amount of edible flesh also increases in these artificially pollinated fruit.

Uses of plant bioregulators

Some plant bioregulators have been used to increase fruit setting. Spraying paclobutrazol at 500 ppm over the entire tree two to four weeks after flowering can increase fruit setting in a similar way to artificial pollination. However, the development of young fruit two to four weeks after full bloom is slower than normal. This slow growth rate will last until 15 weeks after full bloom, after which the growth rate returns to normal. The quality of the harvested fruit does not differ from that of naturally developed fruit. Some durian growers have sprayed NAA, GA_3 and foliar spray fertilisers, such as potassium nitrate, calcium nitrate, calcium-boron, humic acid or even seaweed extract (Florigen®), and claim that these increase fruit setting. However, their results are not clear regarding the amount of chemicals used and the time of application.

Quality Fruit Production

Orchard management in order to produce high yields and good quality fruit follows the basic relationship between sink and source in the durian trees. If the tree develops normally, without any intervention during growth and development, such as water stress (drought or flooding), disease and insect damage, nutrient deficiencies, etc the yield and quality of the harvested fruit will be maintained. The following management techniques can help to achieve this objective:

Fruit thinning

The idea of fruit thinning is to adjust the sink to suit the food source of the tree. This is done at three stages:

First thinning

This should be done by the fourth week after flowering and is done to thin out misshapen or very small fruit, or fruit from different flower flushes. The fruit that should be left on the tree are those that have a uniform shape and large pedicels. There should be roughly 20% more than are expected at harvest time.

Second thinning

This thinning should be done five to eight weeks after flowering. Small and misshapen fruit should be thinned, as well as fruit with a red or brown colour on the spines, which is a sign of abnormal development as a result of leaving too many fruit on the tree. This stage of fruit thinning helps to adjust the balance of sink and source, and aids the development of the aril.

Third thinning

This is done nine to ten weeks after flowering. Small misshapen fruit with a narrow point at the end should be thinned out. This thinning helps the development of the aril, and regulates the size and shape of the fruit left on the tree. The harvested fruit are very valuable by the time they are sold.

Fertilisation

Fertilisation is essential in increasing source capacity as well as efficiently increasing the translocation of nutrients from the leaves to the growing fruit. A high capacity source increases the yield and quality of fruit. The recommended fertiliser is NPK + Mg at a concentration of 12:12:17 + 2 with sulphur as the salt base. This should be applied five to six weeks after flowering. Another fertiliser (NPK at a concentration of 0:0:50) should again be applied seven to eight weeks after flowering. This fertilisation will help to encourage fruit growth, the development of the aril and aril colour.

Control of young shoot development

The sprouting of vegetative shoots during fruit development is not desirable, as the growing shoots will compete with the developing fruit for food. This is one of the major factors leading to immature fruit drop. Managing irrigation from three to eight weeks after flowering by frequently applying small amounts of water can reduce the sprouting of vegetative shoots to a certain degree. Growers should frequently inspect the trees, especially three to eight weeks after flowering, and if vegetative shoots are found, spraying the trees once or twice with potassium nitrate at a ratio of 150 to 300 g/20 litres water is recommended. This has been seen to work well with the Chanee cultivar. For the Mon Thong or Kradum Thong cultivars, spraying the trees with dimethoate at 50 cc/20 litres water to damage the young leaves is recommended, rather than using potassium nitrate (Hiranpradit *et al.*, 1998).

Managing the bearing trees

Durian fruit are rather heavy and large in size, and in some growing areas tearing or breaking of the bearing branches is evident. Tying the bearing branches to the main branches can help reduce the damage by spreading the weight of the fruit among the branches (Plate 9.2).

Weed Control

Weed control is a recurring cost in the durian orchard budget and it has been estimated that, in a commercial orchard in the eastern region of Thailand, weed control may currently cost US$250 to US$300 per hectare per annum. The weed control methods used are usually a combination of mechanical and chemical, depending on the resources the grower has available and the orchard management system adopted. Dry grass and weeds can become a fire hazard during the dry season, but this risk is generally outweighed by its value when used as a mulch to help conserve soil moisture.

Major weed species

Many annual weed species are found in durian orchards in Thailand, and they require regular treatment to keep them under control, often at intervals of a few months. The main weed species include: *Ageratum conyzoide*, *Amaranthus* spp., *Digitaria* spp., *Eleusine indica*, *Euphorbia geniculata* and *Pennisetum* spp. Perennial weeds are a more serious problem and can compete with young durian trees for available water, soil and nutrients. The most prolific of these are *Cyperus rotundus* and *Imperata cylindrica.*

Weed control measures

Mechanical control

Small tractors with attached mowers are normally used in large orchards. Medium-sized and small orchards may use two-wheeled mowers powered by three to five horsepower engines. The cut grass and weeds are used either as a mulch or as compost. During the rainy season, the ground is frequently too wet to permit the use of even light-wheeled machinery and the only practical alternatives to hand-slashing, which is expensive, are the planting of cover crops or the use of herbicides.

Chemical control

The two main types of chemical used as herbicides are paraquat and glyphosate. Paraquat, a non-selective contact herbicide, is used in controlling annual weeds and can be used in the rainy season, although it is not effective against some species. Glyphosate is a non-selective systemic herbicide which is now widely used to control perennial weeds. Its main limitation is that it is not fully effective if it rains within four hours of application. Glyphosate has been found to be particularly useful in the control of *Imperata cylindrica.*

Hand weeding

In very small orchards and home gardens, weeds are controlled by hand-weeding and slashing, which prevent weeds competing with the trees, particularly newly planted ones.

Cover Crops

Newly planted orchards in particular benefit from the establishment of leguminous cover crops between the tree rows; soil erosion is prevented, soil moisture is retained, and the trees benefit from the additional nitrogenous compounds formed due to the activity of the nitrogen-fixing *Rhizobium* and other bacteria. Cover crops also reduce orchard maintenance costs and prevent weed growth; they may also reduce pest infestation. The main cover crops in Thailand are: *Calopogonium mucunoides, Centrosema pubescens, Phaseolus* spp., *Pueraria phaseoloides* and *Stizolobium* spp. (velvet beans), but in other areas *Crotolaria, Vigna* and *Canavalia* spp. are planted (Almeyda *et al.*, 1979). It is essential to keep the base of the trees clear of the more invasive crops by regular slashing.

Intercropping

As an alternative to establishing cover crops in new orchards, the pathways between the trees may be planted with a range of annual food crops for the first few years. This has the advantage of producing a profitable yield before the durian trees reach bearing age. At this early stage in the growth of the orchard, food crops, such as beans, leafy crops and root crops, will not compete with the young durian trees. When it is no longer feasible to grow food crops due to the expansion of the tree canopies, cover crops can be planted.

Mulching

The application of dried grass, weeds and other organic residues as a mulch around the base of both young and mature durian trees is a commonly used means of retaining soil moisture, preventing weed growth, reducing soil temperature and, through incorporation of the decayed residues, contributing to soil fertility. Mulches are usually placed around the base of the tree to a depth of 10 to 15 cm, leaving the area around the base of the trunk clear. Decomposed mulches should be replaced or added to at regular intervals.

Orchard Rehabilitation

Replanting

After 40 to 60 years of productive growth, trees may begin to produce yields which are uneconomical. This is probably due to age, combined with severe pest and disease infestation. Such trees should be removed and replaced with young trees of high yielding cultivars. It is preferable to remove all the roots and most of the soil in the vicinity of the roots, replacing these with good topsoil to which NPK fertiliser has been added. The best period for replanting is the first week of May (early rainy season).

Top-working

It may be necessary to replace an existing non-productive cultivar with an improved one, or to convert an existing cultivar to a newly introduced one. Both these objectives can be achieved by top-working, which has been described by Coronel (1983). The trees are usually cut back in several stages, to within 30 to 50 cm of the level of the soil. The exposed surfaces are painted with a black bituminous or proprietary sealing compound to prevent the entry of water and disease spores, and promote callus formation. Several shoots will develop from latent buds arising from immediately below the cut surface. The weaker shoots are then pruned away, leaving four or five of the most vigorous shoots. When these shoots are about 0.5 cm in diameter or slightly thicker, they may be approach-grafted, cleft-grafted or budded using scion wood from a desired cultivar. It is likely that the growth of scions will be rapid due to the availability of plentiful food reserves, and some fruit may be produced in the third year after budding or grafting.

10

Physiological Disorders and Non-Pathogenic Diseases

Physiological Disorders

Durian suffer from two main physiological disorders. The first disorder, common among many cultivars, is the uneven ripening of the aril of the fruit. This disorder is characterised by the aril remaining hard, firm, dry and tasteless around some seeds, while ripening normally around the other seeds of the same fruit. In some instances the whole aril can be affected.

The other physiological disorder, which is equally common, occurs when the aril around some or all of the seeds becomes mushy and watery, with a flat dull taste. Many growers report that a watery aril is more common in fruit that ripen during the rainy season.

The exact causes of either of these disorders are not known. One hypothesis is that both could be partially related to an imbalance or upset in the normal respiratory processes of the aril. What triggers this, and the mechanisms involved, needs thorough investigation before either disorder can be dealt with. However, proper management of fertilisation, irrigation and drainage can help to reduce the incidence of such disorders.

Nutrient Deficiencies

Durian trees are planted in a wide variety of soils, many of which are lacking in nutrients. Seedlings and newly-planted trees up to the age of two years tend to be more sensitive than adult trees to the lack of macro- and micro-nutrients or to the imbalance of these nutrients in the soil. The symptoms manifested are often compound symptoms caused by a lack of several different nutrients (Plate 10.1). Sometimes the lack of a particular element is discovered once durian are established in soils low in that particular element, or when fertilisation is not properly implemented.

The main effects of a shortage of nutrients are the retardation of growth and development, and an extension of the juvenile period. Trees deficient in nutrients tend to fruit much later, more irregularly and give lower yields as insufficient nutrient reserves cause fruit abortion. It has also been suggested that a lack of nutrients is a factor contributing to the high plant mortality rate observed during the early years of growth.

Symptoms

Symptoms produced by the deficiency of a particular element are best studied by use of nutrient-sand culture experiments. The deficiency symptoms described below are obtained from studies of this type done at the Malaysian Agricultural Research and Development Institute (MARDI) (Lim, 1993).

Nitrogen deficiency

The major symptom is a general light yellowing of the older leaves without appreciable defoliation. The leaf veins also turn yellow.

Phosphorus deficiency

Leaves produced by plants lacking in phosphorus are much smaller than those on plants receiving a complete nutrient solution.

Potassium deficiency

Symptoms first appear on the older leaves with a yellowing of the leaf tip, which extends to the edge of the leaf. Subsequently, the affected areas turn brown and necrotic.

Calcium deficiency

This is identified by the severe stunting of the plant's growth, small leaves, and necrosis of the tips of the intermediate and older leaves.

Magnesium deficiency

This leads to the yellowing of the older leaves, beginning at the intercostal areas near the mid-rib and extending to the periphery. The veins remain green. Affected leaves defoliate.

Sulphur deficiency

Symptoms first appear on the younger leaves which remain small and turn yellow, while the veins remain green. The upper surface of affected leaves assume a convex curve.

Iron deficiency

The younger leaves become chlorotic while the veins remain green and become pronounced.

Manganese deficiency

Plant growth is severely stunted. The older leaves become chlorotic and yellow, and the veins remain green. Affected leaves tend to defoliate.

Copper deficiency

The young terminal leaves become distorted and the edges turn brown. The ensuing dieback of the terminal shoots is followed by a proliferation of lateral shoots which also succumb to dieback.

Treatment

Nutrient deficiencies are best addressed by the use of appropriate fertilisers. Currently, the fertilisation of durian is rather subjective. Fertilisation requirements obviously vary according to soil type. To determine the optimum fertilisation treatments for durian, systematic fertiliser trials, coupled with periodic and reproducible leaf sampling, and foliar and soil analysis procedures, are urgently required.

Generally, in the early stages of crop development, large amounts of nitrogen, phosphorus and potassium are necessary. During the flowering and fruiting years, organic fertilisers are beneficial, and phosphorus and potassium must be replenished in greater quantities. For optimum uptake of both macro- and micro-nutrients, the soil around the tree should be periodically limed.

Nutrient deficiencies may be immediately corrected by repeated applications of a compound foliar fertiliser, or if a certain element is determined to be lacking, by repeated spraying with a foliar fertiliser solution containing that particular nutrient. Foliar fertilisation should be supplementary to regular ground fertilisation.

Drought Stress

Durian trees are susceptible to water stress at all stages of their development, from the seedlings and young field plantings to mature bearing trees. Different clones exhibit different levels of drought tolerance. Clones D 88 and Kradum Thong seem to be the least drought-tolerant among the clones planted in Malaysia and Thailand, respectively. On an adult tree the signs of water stress, as experienced during prolonged periods of drought, include the foliage turning yellow before being shed prematurely, leaving almost bare branches. The bare branches and twigs succumb to the effects of solarisation which in turn paves the way for infection by opportunistic pathogens like *Lasiodiplodia theobromae* and *Colletotrichum gloeosporioides*. Such pathogens cause the secondary dieback of twigs and branches.

Drought stress is increased by low relative humidity, high light intensity, wind and high temperatures. However, most trees will rejuvenate and refoliate with the onset of wet weather, and with adequate fertilisation. In extreme cases, prolonged periods (more than a month) of drought can reduce the vitality of the tree and make them more susceptible to infection by primary pathogens, like *Phytophthora palmivora*, with the onset of wet weather. One orchard in Malacca was virtually destroyed by the combined action of drought and *P. palmivora* disease (Lim, 1993).

A lack of moisture in seedlings and young trees can cause the retardation of growth and development, complete defoliation and, in extreme cases, the death of the plant. In fact, lack of moisture is a major contributing factor responsible for the high mortality rate of newly planted durian trees. The necessary replacement of these trees leads to a corresponding increase in establishment costs. To avoid drought stress, durian plants should be irrigated from the time they are planted out in the field until they begin to bear fruit. Micro-irrigation (drip or trickle) is the most effective form of irrigation to install in durian orchards in Thailand. During prolonged dry periods, young trees should be irrigated at least twice per day and older trees at least once. Additionally, protective light-shades made from coconut fronds, coupled with some form of organic mulch (eg dry grass around the base of the plant), can help to mitigate the effects of drought and solarisation.

Herbicidal Damage

Herbicidal damage is common in durian seedlings and young (one to two year old) trees. It is caused by the excessive use of herbicides, non-uniform application, contaminated or improperly calibrated spray equipment, drift from adjacent areas, use of the wrong herbicide or residual carry-over in the soil.

Paraquat damage

Paraquat is the most commonly used herbicide in durian orchards. It is a non-selective contact bipyridilium herbicide which exhibits no significant soil activity. Careless spraying, especially in strong winds, often causes injury to young trees. Older trees seldom experience such damage, as their canopies are higher up and well out of reach.

Affected leaves turn brown and become necrotic in spots or irregular patches where contacted by spray droplets. Spray drift can be avoided by using a well-calibrated sprayer and attaching a spray shield around the nozzle. Spraying should not be carried out in strong winds. Another precaution is to use separate sprayers for herbicides and other pesticides, to avoid the risk of using herbicide-contaminated equipment.

2,4-D damage

2,4-dichlorophenoxy-acetic acid and its salts (2,4-D) is commonly used in plantation crops but not in durian orchards. Residues of these herbicides can, however, persist in the soils of orchards established in ex-plantation areas. These herbicides are systemic and hormonal in activity. A severe case of 2,4-D injury was reported in a durian nursery where the soil for filling polybags had been taken from nearby oil palm areas where 2,4-D had been used to control broad-leafed shrubs. Around 80% of a recently planted batch of durian seedlings were affected, out of which more than half died and the rest were malformed. Seedlings emerged and appeared normal for a week or two, but as the roots grew and took up more of the residual herbicide, growth and development were affected. The seedlings became stunted and the leaves malformed, failing to unfurl and becoming elongated, twisted or bent (Plate 10.2). Some seedlings outgrew the damage while others died, depending on the amount of herbicide absorbed.

To avoid the effects of herbicide residues, soil used for filling polybags should be taken from topsoil in areas where residual type herbicides have not been used recently, or better still, from virgin areas free from any herbicide residue. Using heat-sterilised soil for polybags can also help to avoid such injury, as most herbicides are not thermostable.

11

Pests and Diseases

Orchard Pests

There are several different insects that can infest durian throughout the various production areas but so far few of them have ever reached epidemic proportions. Pesticides are used in many commercial orchards but smallholding and home garden trees are rarely treated with chemicals. The use of very toxic chemicals, and those with a long-term residual effect should be avoided.

While the pests that generally infest common fruit trees can be satisfactorily controlled with good orchard management combined with some routine insecticidal sprays, limited scientific research has been carried out on the pests that attack durian, and life cycle details and methods of controlling these insects are relatively unknown. Published data on nematode infestation, methods of biological control and integrated pest management is almost non-existent. The following appear to be the most commonly occurring pests in the major production areas:

Leaf-eating caterpillar (Arthisma sp.)

The larvae of this insect feed on young leaves; the caterpillar is brownish-green and similar in colour to the leaves. It can grow up to 8.5 cm long and is found both underneath and on top of the leaf. Pupation occurs in the soil from where the adult butterfly emerges to mate, lay eggs and start the cycle again. If there are not too many caterpillars, they can be manually removed from the leaves. However, if the infestation is severe, spraying with insecticides such as malathion, carbaryl or pyrethrin is recommended.

Stem-boring caterpillar (Zuzera sp.)

The larvae of this insect are similar to those of the fruit fly, except their bodies are red and pink with black heads. The caterpillars burrow their way through durian branches and, by living on the pith, cause the branch to wither and dry up. When this is seen, the affected branch should be cut off and burned. Attempting to kill the caterpillars with insecticide may not be effective as the insect can conceal itself within the branch. The damage may instead be reduced by using a strong-smelling insecticide which will prevent the adult insects from laying eggs.

Flower-eating caterpillar (Pubjemama vesiclora Walk.)

The larvae of this insect feed on young unopened flowers. The full-grown caterpillar is brown and can grow up to 1.5 cm long. Infested flowers are easily identified by the faeces of the caterpillar on the petals of the flower. Affected flowers will gradually dry up and fall off. The caterpillar can best be controlled by thoroughly checking the flowers and manually destroying the insects. If the infestation is likely to be severe, it is recommended that trees be sprayed with systemic insecticides before the flowering period.

Fruit-boring caterpillar (Dichocrocis punctiferalis Guen.)

The larvae of this butterfly are particularly notorious. The caterpillar hides itself beneath the skin of the fruit and feeds on the fruit tissues. The damage is not easily seen and sometimes the caterpillar feeds right to the seeds. The infested fruit can develop to maturity but the infested aril will be damaged and inedible (Plate 11.1). The adult butterfly lays eggs on the surface of the small immature durian fruit and, after hatching, the young larvae bore into the fruit where they feed and develop. The caterpillars (which grow to approximately 3 to 4 cm) are easily seen when the fruit are cut open. The mature caterpillars bore back to the outer part of the fruit and drop to the soil, where they pupate until the adult insects emerge to lay eggs on next season's durian fruit. Systemic insecticides are recommended to kill eggs and young larvae. Spraying should be carried out when the fruit are about the size of an orange and should be stopped at least three weeks before harvesting.

Stem-boring beetle (unidentified species)

The larvae of this beetle are 8 cm long. The larva first attacks the bark of the tree and eats through to the pith of the stem, however, some larvae live on and eat only the bark. When this happens and the bark has been eaten right around the trunk and branches, the tree may die. The first sign of infestation is a light brown ooze on the surface of the bark, together with the faeces of the larva. The best remedy is to scrape away the infested bark and destroy the larva, or using cotton soaked with insecticide, plug the hole that the larva has made in the bark.

Leaf-mining grub (Dactylispa leonardi Rits.)

This grub is the larva of a small beetle in the family Hispidae. The beetle is 4 to 7 mm long. The damage caused by the larvae appears as white patches on the leaves. The grub is generally found in most durian-growing areas of Thailand. In a severe infestation, the best remedy is to spray the trees with an insecticide such as carbaryl.

Leaf-sucking psyllid (Tenaphalara malayensis Crawford)

This is a small insect, 2 to 5 mm long, resembling a small tree hopper or an aphid with strong legs. Both sexes have long tentacles and wings. The young insect excretes a white wax-like substance which covers the insect so that it resembles a mealybug. Both the immature and adult insects are very mobile and cause damage by sucking on the young shoots, causing them to wilt. This insect is also known as jumping plant lice. The most effective remedy is to spray the young shoots with insecticide wherever the damage occurs.

Fruit-sucking tree hopper (Sarritor attenuatus Funkl.)

This insect is pale green with two brown horns between its thorax and abdomen, with a white strip between its two wings. It will jump and fly away when disturbed. This insect sucks the stalk of the young durian fruit causing the fruit to fall. If the attack occurs when the fruit are larger, the fruit may not drop but they do not develop normally and are not marketable. Ants can often be found feeding on the excreta of this insect. If the infestation is epidemic, the trees should be sprayed with systemic insecticides at the very earliest sign of fruit development.

Durian scales (Coccus spp.)

These insects feed on sap from the branches of the durian tree. They are covered with brown scales and are 3 mm in length. The infested branch turns black with the excreta of the insect on which ants can be found feeding. The insect also attacks durian flowers causing them to drop off. Spraying with systemic insecticide is recommended as a remedial measure.

Durian mealybug (Pseudococcus spp.)

These insects are covered with scales and the male insect can fly while the female cannot. The insects mate beneath the bark. After laying her eggs, the female dies and the eggs are left inside the bark where they cause many small spots. Severe infestations result in the decay of the bark, then the branch, and sometimes in the death of the tree. The larvae hatch in the bark; they are very small and cannot be seen without magnification. The insect is woolly in appearance and moves slowly. Damage will become apparent after the third year of planting when the trees have enough bark for the insects to mate and hatch their eggs in. Signs of the disease begin at the base of the tree and spread to the branches. Good orchard management, including proper pruning and regular weeding, can prevent infestations of mealybugs. In cases of severe damage, it is recommended that trees be sprayed with systemic insecticide.

Bark beetle (unidentified species of Bostrichidae, Coleoptera)

This is a dark brown insect, 2 to 3 cm in length. It can cause damage to the durian tree in both its immature and adult stages. The insects feed by piercing the tree at the base of the trunk and sometimes at branch level creating small holes. One can see faeces or ooze coming from these holes. The leaves of the infested trees wilt and drop off, and the tree will eventually die. The best remedy is to scratch off the damaged bark and paint the wound with a fungicide such as cupravit.

Durian diseases

Seedling and root diseases

Dieback and foliar blight

This is caused by the *Phytophthora palmivora* fungus. Loss of seedlings in nurseries can reach alarming levels, in some areas exceeding 50%. Infected nursery soil is the primary source of disease, while overcrowding and excessive watering of shaded plants will provoke the rapid spread of disease. The infection usually begins at the stem and extends to the leaves and roots. When the trunk and branches have been girdled by the infection, seedling dieback ensues. The disease causes rootlets and tap roots to rot, and also prevents further rootlets forming. When the infection reaches the leaves, small, off-coloured, water-soaked spots are seen (Plate 11.2). These spots darken and coalesce into larger necrotic patches (Plate 11.3). Infected leaves droop and fall prematurely. If the infection continues, the whole foliage becomes blighted and the seedling dies.

Disease management

The disease can be minimised by spacing the seedlings more widely, maintaining good drainage, and avoiding excessive and careless watering. Infected seedlings should be removed. The use of sterilised soil in the polybags used to raise seedlings will help to avoid the disease. To halt the spread of disease in nurseries, periodic sequential spraying with systemic fungicides, such as metalaxyl, milfuram, orfurace, cyprofuram, oxadixyl or fosetyl aluminium, is recommended. Products which combine different fungicides, such as metalaxyl and mancozeb, or orfurace and mancozeb, will help to avoid fungicide resistance. Spraying should be thorough, with a sticker-spreader added if desired. The above fungicides can also be used as soil drenches.

Pythium root rot

Root rot in durian trees is caused by *Pythium vexans* de Bary. Its synonyms include *Py. allantocladdon* Sideris, *Py. euthyhyphon* Sideris, *Py. complectans* Braun, *Py. polychadon* Sideris and *Py. piperinum* Dastur. The fungus is homothallic, ie capable of producing oospores on its own. Rather than attacking the main trunk as *P. palmivora* does, *Pythium* attacks the roots of the tree and may prove fatal.

Pythium root rot is not confined only to full-grown durian trees but is also common in seedling nurseries where 4 to 12 week old polybag seedlings have succumbed to the disease. Such infected seedlings usually die. Scorching seedlings by improper and careless fertilisation has been observed to increase the risk of *Pythium* infection. In a typical case, the lower lateral roots on one side of the tree become necrotic from the tips. This spreads through the cortex and the wood to the tap root, which begins to decay downwards. Externally the roots appear normal, but internally the cortex becomes dark brown and the wood appears pinkish with flecks of brown. In more severe instances, the rot extends from the tap root upwards to girdle the cortical tissue at the collar (Plate 11.4). Once the collar has been infected, above-ground symptoms ensue in the form of branch dieback. New shoots are often produced beneath the dieback if the roots are not completely damaged. Once it has been completely girdled, the tree will eventually die.

Infection in seedlings is first observed in the flaccid and limp appearance of the leaves. Brown patches appear at the leaf base. Within a few days all the leaves will have wilted and the seedlings will die. Infected seedlings show extensive decay of the roots, tap root and cortical tissues, vascular bundles, and the pith of the stem near the base of the buds. This rot girdles the stem and interferes with the uptake of nutrients and water which causes the above-ground foliage to wilt and eventually die. Infection can begin from the roots or from the buds.

Disease management

Several remedial measures are recommended. These include the uprooting and burning of dead infected trees, the construction of trenches around infected trees, and the inspection of adjacent trees. Mildly infected trees should have root surgery and exposed root surfaces should be painted with a wound dressing. However, the most effective treatment nowadays is the use of systemic fungicide soil drenches, such as metalaxyl, fosetyl aluminium, propamocarb hydrochloride, or etridiazole. In the nursery, *Pythium* infections can be prevented by the use of sterilised soil when growing seedlings in polybags. Proper field drainage, careful fertilisation, and the avoidance of excessive watering can also help in thwarting the disease. As the fungus is also vulnerable to certain soil fungi like *Trichoderma* spp., some durian growers in eastern Thailand have begun using the *Trichoderma* fungus as a means of biological control. Another possibility is the use of resistant cultivars or root stocks, but more studies must be carried out in this area before this may be practised widely.

White root disease

This disease, which is soil-borne, can be a particular problem in those durian orchards established on ex-rubber or ex-cassava land. An infected tree usually dies if not treated in the early stages of infection. Trees of any cultivar and age can succumb to infection, but the disease is most devastating if it occurs in the first six years after planting.

The pathogen of this root disease is *Rigidoporus lignosus* (Klotzsch) Imazeki, (synonyms: *Fomes lignosus, Leptoporus lignosus*). It is a basidiomycete fungus of the family Polyporaceae. The infected trees exhibit easily diagnosed above-ground symptoms. The foliage wilts and changes colour from green to yellow to brown, and finally shrivels. The tree usually dies shortly after displaying these symptoms. Below ground, the diseased roots are engulfed by the white rhizomorph growths of the fungus. Rhizomorphs of the fungus establish themselves all over the root system, including the tap root (Plate 11.5).

Disease management

For this root disease, 'prevention is better than cure'; the best method of control is avoidance of the disease. This can be done by thoroughly clearing land in preparation for growing durian. All stumps, roots, and woody debris of the felled rubber, cassava, or jungle trees should be uprooted, removed, and burned. Dead, diseased trees should be felled, completely removed with all their roots, and burned. The vacant holes should be drenched with a fungicide solution, such as tridemorph (Calixin®), triadimefen (Bayleton®), propiconazole, or another triazole. A further treatment, dusting the vacant holes with sulphur before replanting, is also advisable. The roots and collars of adjacent trees should be inspected and, if found to be infected, should be drenched with any of the above fungicides and treated with sulphur. The construction of isolation trenches around diseased trees can also help curb the spread of the disease.

Xiphinema nematode

Xiphinema (otherwise known as the dagger nematode) is an ectoparasitic nematode. It was first reported infecting durian roots by Razak (1988). The nematodes feed on the roots and damage them, which further aggravates trees suffering from stresses such as drought or lack of nutrients.

The nematode feeds on the apical tip of the feeder roots, retarding their growth and causing the root tip to swell. In consequence, feeder roots are stunted in growth, being less than one third the length of the normal healthy feeder roots. Affected roots resemble a herring bone with offshoots of short stubby rootlets (Plate 11.6a and b). Some of these stubby rootlets degenerate, dry up and become friable. It is difficult to differentiate above-ground symptoms from those caused by drought and lack of nutrients.

Disease management

As the economic impact of this nematode upon durian production is difficult to assess, control measures may not be warranted unless there is an extremely high nematode population in the durian rhizosphere. If necessary, nematicides, along with proper irrigation and fertilisation, can help to lower the nematode population in the soil and revitalise the root system (Lim, 1993).

Trunk and branch diseases

Durian patch canker

Patch canker (or stem canker) together with root rot, leaf blight, and dieback of both seedlings and mature trees are caused by the notorious fungus, *P. palmivora.* This fungus is thought to be the most devastating pathogen to affect durian trees and also causes serious rot in durian fruit.

The first sign of patch canker is the appearance of wet lesions on the bark surface, usually near the crotch region (Plate 11.7a and b), although other parts of the trunk and main branches are equally susceptible. Bark necrosis is accompanied by discolouration of the bark and exudation of a reddish-brown resinous substance. Beneath the affected bark, the cortical tissues and wood appear a dull reddish-brown colour, sometimes with dark claret or purple specks. In contrast, healthy wood is cream-coloured. The wood lesion has a well-defined, but irregular, advancing margin. The infection usually extends deep into the wood and when the infected wood is knocked with a hard object, it sounds hollow. As the lesion expands and girdles the tree, some branches begin to exhibit signs of stress. The foliage wilts and begins to dry up, and branch dieback follows. As the disease progresses, the entire tree may defoliate and die. Infected bark usually becomes riddled with pin-holes made by the pin-hole boring beetle, *Platypus cupulatus* Chap. Underground, the lateral and feeder roots may also be infected, and appear necrotic and decayed. When the decay extends to the main tap root, the whole tree dies. Infections usually begin at the stem, but in some cases may start at the roots and extend upwards along the trunk.

Disease management

Due to the seriousness of this disease, a multi-pronged approach to control is most effective. This should combine environmental, chemical and biological measures, and the use of resistant cultivars.

Environmental Patch canker is spread more easily in a moist humid environment. The use of wider spaces between the trees, the removal of thick weed undergrowth, and the pruning of low-hanging branches will improve air circulation and lower the humidity around the trees. Budded trees should be trained or pruned to allow a straight trunk at least two metres from ground level, as they have a propensity to branch close to the ground, increasing the possibility of infection sites at the crotch. All exposed surfaces should be dressed with an appropriate fungicide, such as captafol, any copper fungicide or fentin acetate. Healthy trees may be maintained and patch canker avoided by adequate fertilisation, irrigation during dry periods, and regular spraying with a mixture of an insecticide, a fungicide and a foliar fertiliser to control other debilitating foliar diseases and pests. Any environmental stress, such as that experienced during prolonged dry periods, can aggravate infection by the fungus and hasten the death of the tree.

Dead diseased trees should be uprooted, removed and burned. The vacant hole should be drenched with an appropriate fungicide and also treated with sulphur to encourage the proliferation of antagonistic soil microflora. In disease-prone areas, monthly or bimonthly checks for the disease should be made. Treatment should be carried out at the first sign of fresh infection.

Chemical The diseased bark should be scraped away and the infected wood should be painted with a fungicide. There is a wide choice of chemicals available for use including protectants, such as any copper fungicide, captafol, fentin acetate, maneb, etridiazole and mancozeb, as well as systemic fungicides, which have both protective and therapeutic properties. The latter group includes: metalaxyl, fosetyl aluminium, ofurace, oxadixyl, milfuram and cyprofuram. It is recommended that systemic fungicides be used in combination with a protectant (such as mancozeb or maneb), or in a sequential application with captafol or etridiazole, to thwart the development of resistance against the fungicide.

Another chemical approach is to drench the soil periodically around infected and adjacent trees with any of the above systemic fungicides. Granules of such fungicides, if available, can also be used instead of soil drenching. Weeds and leaf litter beneath the tree should be removed before drenching or the use of granules. Painting the trunk with fungicide, combined with soil drenching or granular application, gives even better control. Injecting the trunks of diseased trees with metalaxyl, fosetyl aluminium, or phosphorus acid has also been reported to effectively control the disease (Tai, 1971). The control of insect vectors and snails with appropriate insecticides and molluscicides will also help curb the spread of the disease throughout the tree.

Biological Antagonistic soil fungi are a potential means of controlling the disease. Studies show that the sporangia and chlamydospores of *P. palmivora* can be aggressively parasitised and killed by an antagonistic soil fungus, *Gliocladium roseum* (Lim and Chan, 1986). However, more studies are required before this can become a viable field control method.

Resistant cultivars Resistant cultivars are an alternative method of control. With further research regarding the clonal resistance of durian to the disease, this method could become a more common and viable option.

Pink disease

This disease attacks the twigs and smaller branches of durian trees, causing wilting and dieback of pockets of associated foliage. The pathogen of pink disease is the basidiomycete fungus *Corticium salmonicolor* Berk. & Br. of the Corticiaceae family. Mature trees over four years old, with well-developed bushy canopies, are prone to this disease. It is commonly observed in durian orchards in areas of heavy rainfall, in particular inland regions. It has been recorded in Sabah, Sarawak, Peninsular Malaysia and southern Thailand.

The disease first appears as silky white mycelial threads on the bark of small branches or twigs (Plate 11.8). Under moist conditions, the enveloping threads become pink in colour. As the bark dies, the pink mass of mycelial threads form a rough pink encrustation on the infected bark. This is the mature stage and may take several weeks or months to develop. The crust is often broken up by irregular cracks. The infection is identified by the pinkish-white threads covering the shoot and can spread quickly. The infected wood turns dark brown, leaves yellow and drop off, and the branch dies. Although the disease results in the wilting and death of foliage on the infected twig or branch, the whole tree is not killed.

Disease management

Success in controlling pink disease depends on correct and early diagnosis of the disease, followed by prompt control measures. This disease can be confused with the velvet fungus and hence mistreated.

Environmental Wider tree spacing and well-ventilated canopies will reduce the incidence of pink disease. Dead diseased twigs should be manually removed and the exposed site treated with a wound dressing, such as captafol or another appropriate fungicide. The removed diseased foliage should be burned.

Chemical Prompt application of the following fungicides by topical spray or by brushing on, can afford more effective control: triadimefen (Bayleton®), tridemorph (Calixin®), oxycarboxin (Plantvax®), flusilasol (Punch®), or other systemic triazole compounds. More common chemicals, such as copper hydroxide (Kocide®), copper oxide (Perenox®), copper oxychloride (Coprantol®, Zeppercide®), and Bordeaux Mixture (copper sulphate : calcium oxide : water = 1:2:10) can also be used, but require higher rates and more frequent applications.

Velvet felt fungus infestation

This fungus has been observed on durian in most production areas of Malaysia and some parts of Thailand. The stroma of the fungus *Septobasidium* sp., a basidiomycete fungus, appears as a spongy violet mat. It is commonly found encircling portions of twigs and small branches of durian, and other tropical fruit trees (Plate 11.9). Infestation is prevalent during rainy periods and is usually on trees with bushy canopies which have not been sprayed with pesticides for some time.

This infestation is often misidentified as pink disease but, unlike pink disease, which kills the foliage of the affected twig, velvet felt fungus is not pathogenic to durian. It does not cause any direct injury to the tree and, in fact, can be construed as beneficial, as it parasitises certain scale insects which infest the durian tree. However, it does detract from the tree's normal appearance.

The fungus is easily diagnosed – its spongy, felt-like, violet or violet-grey growth is very visible on twigs and small branches. The advancing margin of the fungus is white and delicately fan-shaped. This encrustation can be easily removed as the fungus stroma do not penetrate the bark.

Disease management

The infestation does not harm the tree, except for giving an unsightly appearance to the branches and twigs, and hence warrants no control. However, controlling the scale insects with an appropriate insecticide, such as malathion or white oil, is justifiable and this will reduce infestation by velvet felt fungus.

Lichen infestation

Lichens frequently occur on the surfaces of the leaves, branches and twigs, and on the trunk bark of durian trees. Lichens are termed corticolous when they occur on living bark, lignicolous when they occur on non-living bark, exposed or cut wood, and folicolous when they grow on leaves. They are often misconstrued as injurious parasitic growths. Lichens do not parasitise the tree as they obtain their nutrition from the air and from dead organic matter. They cause some mechanical interference to respiration and transpiration processes when lenticels of the bark and stomates of the leaves are blocked by their growth. In addition, folicolous lichens shade leaves therefore reducing the photosynthesis area and detracting from the well-kept appearance of the tree.

Lichens are extremely sensitive to copper or oil sprays and can be easily eradicated by frequent spraying with any copper fungicide or oil emulsion. Thus, fungicides formulated in emulsifiable concentrates may be used. To achieve maximum effect, spraying should be done when the lichens are dry.

Moss infestation

This is usually found on the bark of the trunk and the large spreading branches where it forms green velvety cushions or extensive mats. Mosses are also found on durian leaves. They tend to proliferate on the upper side of sloping branches that face the prevailing winds and receive the greatest amount of rain and run-off. In particular, neglected old trees with sparse foliage tend to harbour the growth of mosses.

Like lichens, mosses do not injure the tree but rather manufacture their own food by photosynthesis. However, they do block bark lenticels and stomates when they occur on leaves, interfering with respiration and transpiration processes. They may also interfere with flowering and fruiting since durian bear flowers and fruit directly from the bark. Indirectly, they also encourage the spread of disease as they absorb and retain moisture for long periods, creating a moist microenvironment conducive to the spread of disease and the growth of epiphytes.

A well-maintained tree is unlikely to encounter a moss problem. If treatment is necessary, manual removal or spot spraying on cushions or tufts with a sodium chloride solution will eradicate the mosses. Mosses are also extremely sensitive to salt water which damages the delicate lamina cells of the moss (Lim, 1993).

Foliar diseases

Rhizoctonia foliar blight

This disease infects the foliage of both seedlings and adult durian trees, and is caused by the fungus *Rhizoctonia solani* Kuhn, the anamorphic state of the teleomorph *Thanatephorus cucumeris* (Frank) Donk. (Lim *et al.* 1987). In the nursery it can be lethal to seedlings, killing approximately 40% to 50%. On mature trees it causes the leaves to die, dieback of twigs and small branches, defoliation, and subsequently a decrease in yield.

The disease begins as small water-soaked spots on the leaves which coalesce to form larger, irregular, water-soaked patches. These lesions subsequently dry up into light brown, irregular patches with dark brown margins that deform and curl the leaves (Plate 11.10a and b). Affected twigs appear shrivelled, desiccated and necrotic. Under warm moist conditions, buff-coloured, mycelial threads of the fungus appear on the diseased tissue and spread to adjacent leaves. Many tiny, light brown sclerotic patches frequently appear on the diseased tissue in these conditions. Infected leaves fall off easily and in severe cases of infection, portions of the canopy may become devoid of leaves. This drastically reduces the photosynthetic area and has a negative impact on flowering and fruiting. Infected seedlings and twigs shed their leaves prematurely and subsequently dieback. In mature trees, this can result in a thinned-out, sparse, irregular canopy.

Disease management

At the seedling stage, the disease can be avoided by regular but not excessive watering and by spacing the plants widely, since overcrowding and excessive watering tend to increase spread of the disease. The disease can be effectively controlled by several applications of any of the following fungicides: pencycuron (Monceren®), benomyl (Benlate®), carbendazim (Derosal®, Bavistin®, Delsene X®), thiophanate methyl (Topsin M®), flutolanil (Moncut®), and triadimefen (Bayleton®). These can be applied singly, sequentially, or in a soil drench programme. Excellent results are also obtained when the fungicides are injected directly into the trunk. Removing and disposing of diseased twigs and foliage, as well as reducing any undergrowth beneath the tree, will also help prevent the disease.

Phomopsis leaf spot

This leaf spot disease is caused by the fungus *Phomopsis durionis* Syd. Infected leaves tend to abort prematurely. Seedlings are particularly susceptible to damage by this fungus, being completely defoliated in cases of severe infection. Once defoliated, the bare shoots succumb much more easily to sunburn and are open to secondary infection by fungi such as *Lasiodiplodia theobromae* and *Colletotrichum gloeosporioides*. The disease is not such a problem in mature trees with well-developed lush canopies. Early signs of the disease are tiny spots on the leaves, each with a yellow halo. The spots enlarge to become circular or oval, approximately 1 mm in diameter. They are ash-white to light brown in colour, with necrotic, dark brown margins surrounded by distinctive yellow haloes. The spots occur randomly all over the lamina on the upper surface of the leaf. On the reverse side of the leaf, the spots are difficult to see as they are masked by the peltate leaf scales. As the fungus matures, the fruiting bodies of the fungus can be seen as minute black spots in the centre of the lesion (Plate 11.11).

Disease management

Very little is known about the epidemiology of this leaf spot. There may be an association between damage caused by insects and fungal infection. In mature trees with lush foliage, no treatment is necessary but intervention is important for seedlings. The disease can be controlled with systemic fungicides, such as benomyl, carbendazim, thiabendazole, or thiophanate methyl, mixed with protectant fungicides, such as mancozeb, propineb, or chlorothalonil. Appropriate insecticides should be used to control leaf-sucking insects. Some of the triazole systemic fungicides, such as propiconazole, flusilazol and flutriafol, are also fungitoxic. Prematurely aborted leaves should be removed from the vicinity of the seedlings and burned to eliminate sources of inoculum.

Leaf anthracnose

This disease is commonly found on the leaves of seedlings and debilitated trees. Infected leaves may prematurely abort. The anthracnose pathogen is *C. gloeosporioides* Penz., the anamorphic state of *Glomerella cingulata* (Stonem.) Spauld & Schrenk. The pathogen is also associated with a secondary dieback of durian twigs and branches. The incidence of dieback may be related to a combination of meteorological, environmental and nutritional factors, which weaken the trees, predisposing them to infection by *C. gloeosporioides*. The disease occurs on durian leaves as discrete, round, necrotic leaf spots or irregularly-shaped necrotic patches usually near the margin or leaf tip. The leaf spots have pale greyish-brown centres, with more or less concentric rings of tiny black spots and dark brown, necrotic margins usually surrounded by a pale yellow halo (Plate 11.12). The irregularly-shaped patches are characterised by dark brown, zonate lines near the advancing margins. Infected leaves on seedlings or debilitated trees may be prematurely shed.

Disease management

The recommended method of control for anthracnose leaf spot in the nursery is to apply several foliar sprays. A mixture of fungicides, such as benomyl, thiophanate methyl, or carbendazim, in combination with chlorothalonil, propineb, metiram, mancozeb, or maneb is suitable. The addition of an insecticide and foliar fertiliser to the fungicide is also recommended to revitalise debilitated trees. This is also useful for diseased seedlings and mature trees.

Cephaleuros algal rust

This rust is a green alga, *Cephaleuros virescens* Kunze, and it causes spots on durian leaves. It weakens the tree by reducing the leaves' photosynthetic ability. However, the disease is not difficult to control. The damage caused by the disease is aggravated under environmental stresses, such as poor soil conditions, thick undergrowth, overcrowding, and insect and mite infestation.

Signs of the disease are rust-coloured velutinous spots varying from 3 to 5 mm in diameter, which are usually found on the upper surface of the leaves and on young twigs or small branches. Such spots may merge to form irregularly-shaped larger patches. The infestation may cause the bark on twigs to crack, opening portals for secondary infections by microorganisms. At a later stage, the orange spots may turn greyish-green, as the infection moves into another phase of its life cycle. When the spot is scraped away, a thin, greyish-white to dark coloured, necrotic crust is left on the lamina, which reduces the normal photosynthetic area of the leaf.

Disease management

Algal rust infections are most serious when the tree is unhealthy. Therefore a logical way to control the disease is to revitalise the tree by properly fertilising and irrigating, alleviating nutritional deficiencies, reducing the humidity around the trees, and increasing air circulation in the canopy (by removing any undergrowth and spacing the trees more widely). Stubborn infections can be alleviated by spraying several times with an algicide or fungicide, such as fentin acetate, benomyl, thiram or any copper fungicide (Lim, 1993).

Fruit diseases

P. palmivora fruit rot

The familiar fungus, *P. palmivora*, also causes a serious fruit rot in durian. Affected fruit completely lose their marketability. During the rainy season, as much as 30% of an orchard's fruit can be affected. The first sign of this disease is hydrotic patches on the surface of the fruit. These turn brown and become darker over time. Under moist conditions, a white bloom of cottony mycelia and sporangia forms on the necrotic lesion (Plate 11.13). Internally, the rot can extend to the arillate seeds.

Disease management

Directly spraying the developing fruit with any of the fungicides discussed for patch canker control is one avenue of control. However, it is important that fungicides do not leave any toxic residues on the edible fruit. The height of the fruit on the tree makes it sometimes difficult to spray all of the fruit. This necessitates expensive spraying equipment, such as a high-pressure spray gun system. Injecting any of the aforementioned systemic fungicides straight into the trunk offers an effective alternative.

Sclerotium fruit rot

The causal fungus of this disease is *Sclerotium rolfsii* Sacc., the sclerotial state of *Corticium rolfsii* Curzi. In a durian orchard, this fruit rot can be easily identified by white fan-shaped tufts of thick mycelial strands growing on a water-soaked, brown, necrotic lesion on the fruit (Plate 11.14a). It is commonly found on fruit which has fallen to the ground. Internally the decay extends into the seed cavity, rotting the seeds and discolouring the cotyledons (Plate 11.14b).

Disease management

To avoid *Sclerotium* rot, fallen fruit should be picked up and discarded daily. The base of the durian tree should be kept free from undergrowth while fruit is ripening, as any undergrowth will create a warm moist environment ideal for the spread of infection by soil fungi.

Post-harvest fruit rots

Post-harvest fruit rots take various forms depending on the nature of the causal agent. Generally the lesions produced on the fruit consist of irregular necrotic patches of various shades of brown which darken over time. In the case of those associated with *Rhizopus* (Plate 11.15), *Mucor* or *Lasiodiplodia* (Plate 11.16), the lesions are soft when pressed and the necrosis often extends into the inner wall of the rind. Lesions associated with *Phyllosticta, Colletotrichum* and *Curvularia* are firm to the touch and superficial (Plate 11.17). Lesions associated with *Fusarium* are water-soaked, soft and pulpy, and are usually covered with the white or salmon-coloured mycelia and spores of the fungus.

Disease management

Post-harvest fruit rots can be minimised by more frequent fruit picking (two or three times a day), and by brushing off soil and leaf debris stuck to the rind or thorns of detached fruit. Reducing the impact of the fruit falling to the ground by constructing nylon or wire nets beneath the trees during the fruit ripening season also helps. This will reduce bruising and injury to the fruit and also keeps the fruit off the ground (Lim, 1993).

Part 3

Post-Harvest Management, Costs and Marketing

12

Costs and Returns of Durian Production

Very little information is available about investment in tropical fruits, especially durian. It should be remembered that many tropical fruits are grown in smallholdings, or 'dusun', as in Malaysia, and in mixed orchards as in Thailand. Some of the methods of estimating the costs and returns of durian production may not be 100% correct. There are many uncertainties, including variable climatic conditions causing drought, flood and wind damage, and pests and disease. These factors vary each year. Because of these uncertainties, a budget based on estimates can only be a guide. The decision-maker must consider each factor separately, select a high and low option for each important variable and then input this data into the completed model. Risk can then be estimated from the sensitivity analysis performed on the model.

The uncertainty regarding the profitability of irrigating durian orchards illustrates that the ideal production system has yet to be defined. Initially, it was thought that irrigation was necessary and profitable for commercial durian plantations, but not all the research trials support this assumption. Irrigation is a major capital cost and, if it can be avoided, a major initial saving. It is likely that the debate on the merits of irrigation will go on for many years.

Plantation crops are more complex to analyse than annual crops and livestock enterprises. Simple gross margin budgets are meaningless as durian orchards may take from six to seven years before income exceeds costs.

Break-even price

This is the long-term price required to cover costs, including a return on capital. The return on capital is a real rate of interest, which means that the effect of inflation has been removed. Although somewhat simplistic, the real rate of interest is the charged rate (the nominal rate) less the inflation rate. If the interest rate is 13% and the inflation rate is 8%, the real rate of interest is 5%. A 5% real rate of interest is often selected for agricultural investments.

Internal rate of return

The internal rate of return is the maximum real rate of interest that a borrower can pay on money invested in a project and break even. Constant dollar values are used (ie adjusted for inflation) which means that a dollar will purchase the same amount of goods in any year during the life of the project.

Annual costs

Annual costs include items such as fertiliser, herbicide, insecticide, fuel and labour. It appears that these costs vary greatly from orchard to orchard, and that the model (Table 12.2) is sensitive to changes in items such as fertiliser. Labour costs vary from year to year and between locations. These labour costs include harvesting.

Returns

Yield and price determine the returns. The bearing life of a grafted orchard is usually 45 years. However, soil erosion and wind could prove to be major problems and in this case a greater allowance should be made in the model for tree replacement. It is quite possible that new and better varieties could displace older varieties. Therefore, the models were run for varying periods.

The yield per tree depends on many factors including season, tree density and age. Table 12.1 provides the estimated yield of fruit per tree by age range. Exactly how closer plantings influence the yield per tree will become evident over the next few years as orchards mature and this information becomes available.

Table 12.1: Expected yield (kg/tree) of four durian cultivars in Thailand at different bearing ages (Department of Agricultural Extension, Thailand)

	Bearing age		
Cultivars	*4–6 years*	*7–20 years*	*> 20 years*
Chanee	40	116	181
Kradum Thong	54	102	81
Mon Thong	35	98	215
Kan Yao	21	46	184

Durian Production Costs

There is an increasing trend towards using chemical fertilisers in durian orchards in Thailand. Data from the Department of Agricultural Extension indicates that in 1996, 812 688 tonnes of fertiliser were used on perennial fruit trees. This usage increased to 853 940 tonnes in 1997 and 897 325 tonnes in 1998. This demonstrates that the cost of fertiliser is a high percentage of the total production costs. The calculated cost of fertiliser for durian orchards is listed in Table 12.2.

Table 12.2: Cost of fertiliser in durian production (US$/ha) in Thailand

	Chanee		Mon Thong	
Age of tree (years)	Chemical fertiliser	Organic manure	Chemical fertiliser	Organic manure
1	35.9	9.8	81.8	6.2
2	62.8	7.3	111.9	7.2
3	96.8	3.2	136.6	2.3
4	141.9	5.1	155.1	12.1
5	169.7	4.5	188.6	59.2
6	386.5	1.6	237.6	69.0
7	262.5	26.4	256.0	20.3
8–10	320.1	2.7	339.0	12.0
11–20	411.0	4.6	482.0	4.9

In addition to fertilisers, other agricultural chemicals, such as herbicide, insecticide and fungicide, are widely used in durian orchards. Despite the cost of these chemicals there is an increasing trend towards using them in durian production in Thailand (Table 12.3).

Table 12.3: Estimated cost of agricultural chemicals in durian production (US$/ha) in Thailand

Age of tree (years)	Chanee	Mon Thong
1	75.1	182.3
2	106.0	138.8
3	122.4	149.8
4	130.6	221.1
5	208.2	252.2
6	309.7	363.6
7	314.8	607.4
8–10	498.7	743.8
11–20	733.8	1078.6

At present, there is a tendency to use machinery and modern equipment in the production of durian. This is especially evident in larger orchards where the use of machinery reduces labour costs. The problem of finding both casual and skilled labourers to work in the orchards is becoming serious, as the majority of workers today prefer to work in the industrial sector. There is also the problem of limited cultivated land, and this forces durian growers to do whatever possible to increase productivity.

Durian Production Returns

The data calculated by the Department of Agricultural Extension of Thailand in 1994 on the costs and returns of producing Mon Thong durian in Phang-Nga province, southern Thailand showed a net income of US$2 995 per hectare of planting (Table 12.5). This profit is quite satisfactory as far as the grower is concerned, as many growers in that area grow not only durian but practise mixed cropping in the same manner as 'dusun' orchards in Malaysia. Mixed cropping ensures extended income from different crops. Whether mixed crop orchards

***Table 12.4:* Production costs in cultivating Mon Thong durian (US$/ha) in Chumphon Province, Thailand during the year 1992/1993 (Department of Agricultural Extension, Thailand)**

Items	Year 1	2	3	4	5	6	7	8	9	10	11	12	13	14	15	16	17	18	19	20
1) Fixed costs	261.9	694.1	406.5	578.3	443.6	513.7	513.7	513.7	513.7	513.7	349.4	349.4	349.4	349.4	349.4	485.4	485.4	485.4	485.4	485.4
– Land tax	1.6	1.6	1.6	1.6	1.6	1.6	1.6	1.6	1.6	1.6	1.6	1.6	1.6	1.6	1.6	1.6	1.6	1.6	1.6	1.6
– Depreciation of machinery	260.3	692.5	404.9	576.7	442.0	512.1	512.1	512.1	512.1	347.8	347.8	347.8	347.8	347.8	347.8	483.8	483.8	483.8	483.8	483.8
2) Variable costs	1211.5	1361.9	1011.8	1061.0	1345.4	1804.1	1804.1	1804.1	1804.1	1804.1	1725.9	1725.9	1725.9	1725.9	1725.9	1272.3	1272.3	1272.3	1272.3	1272.3
Labour cost in																				
– Land preparation (till harvest)	502.8	467.5	316.5	450.7	460.8	587.4	587.4	587.4	587.4	587.4	591.7	591.7	591.7	591.7	591.7	452.6	452.6	452.6	452.6	452.6
– Planting	346.0	0	0	0	0	0	0	0	0	0	0	0	0	0	0	0	0	0	0	0
– Cultural practices	362.7	894.4	695.3	610.3	884.6	1167.4	1167.4	1167.4	1167.4	1167.4	1084.9	1084.9	1084.9	1084.9	1084.9	770.4	770.4	770.4	770.4	770.4
– Harvesting	0	0	0	0	0	49.3	49.3	49.3	49.3	49.3	49.3	49.3	49.3	49.3	49.3	49.3	49.3	49.3	49.3	49.3
3) Cost of materials	450.8	507.8	482.8	409.3	607.9	854.7	854.7	854.7	854.7	854.7	872.1	872.1	872.1	872.1	872.1	480.5	480.5	480.5	480.5	480.5
– Plant materials	104.3	0	0	0	0	0	0	0	0	0	0	0	0	0	0	0	0	0	0	0
– Fertilisers and lime	59.9	203.1	301.3	175.7	365.8	382.9	382.9	382.9	382.9	382.9	461.1	461.1	461.1	461.1	461.1	163.4	163.4	163.4	163.4	163.4
– Surfactant	16.6	64.3	28.9	44.8	47.1	35.0	35.0	35.0	35.0	35.0	41.8	41.8	41.8	41.8	41.8	43.8	43.8	43.8	43.8	43.8
– Fuel	47.4	53.8	48.9	55.8	39.0	71.8	71.8	71.8	71.8	71.8	80.3	80.3	80.3	80.3	80.3	47.3	47.3	47.3	47.3	47.3
– Pesticides, herbicides etc.	208.4	184.8	102.3	129.8	146.3	355.2	355.2	355.2	355.2	355.2	257.2	257.2	257.2	257.2	257.2	207.4	207.4	207.4	207.4	207.4
– others	14.2	1.8	1.4	3.2	9.7	9.8	9.8	9.8	9.8	9.8	31.7	31.7	31.7	31.7	31.7	18.6	18.6	18.6	18.6	18.6
4) Other expenses	257.9	386.6	212.5	200.9	276.6	361.9	361.9	361.9	361.9	361.9	262.2	262.2	262.2	262.2	262.2	339.2	339.2	339.2	339.2	339.2
– Repairing equipment	100.0	166.3	60.6	25.3	84.9	113.6	113.6	113.6	113.6	113.6	39.8	39.8	39.8	39.8	39.8	150.9	150.9	150.9	150.9	150.9
– Opportunity cost	157.9	220.3	151.9	175.6	191.7	248.3	248.3	248.3	248.3	248.3	222.4	222.4	222.4	222.4	222.4	188.3	188.3	188.3	188.3	188.3
5) Total costs (average per hectare)	2182.1	2950.4	2113.6	2656.0	2673.5	3534.4	3534.4	3534.4	3534.4	3534.4	3723.7	3723.7	3723.7	3723.7	3723.7	2577.4	2577.4	2577.4	2577.4	2577.4

are suitable for durian production in all areas of Thailand is questionable because not only the kinds of crop grown, but also the market, education, culture and customs of the local growers, can contribute to the success of the system involved. This is a complex system and needs further research.

***Table 12.5:* Costs and returns of producing Mon Thong durian in Phang-Nga Province, Thailand in 1994. (The data are expressed as US$/ha and are from an average of tree ages) (Department of Agricultural Extension, Thailand)**

Items	*Cash*	*Non-Cash*	*Total*
A) Variable costs	680.6	326.3	1 006.9
1) Labour	139.6	312.3	451.9
- General maintenance	39.5	138.0	177.5
- Fertilisation	0.9	17.0	17.9
- Herbicide application	4.6	9.8	14.4
- Pesticide application	14.4	28.2	42.6
- Weeding	17.2	17.9	35.1
- Pruning and tying branches	0.2	46.2	46.4
- Watering	2.3	18.8	21.1
- Harvesting, packing and transporting	60.5	36.4	96.9
2) Materials	445.4	–	445.4
- Planting materials	–	–	–
- Manure	–	–	–
- Chemical fertiliser	186.6	–	186.6
- Herbicide	36.0	–	36.0
- Pesticide	184.4	–	184.4
- Fuel and oil	26.7	–	26.7
- Electricity	1.4	–	1.4
- Other agricultural tools	10.3	–	10.3
3) Miscellaneous	95.6	14.0	109.6
- Interest and opportunity cost	66.1	14.0	80.1
- Repairs to agricultural machinery	29.5	–	29.5
B) Fixed costs	422.3	479.3	901.6
- Expenses before fruit bearing	412.3	187.1	599.4
- Land tax and rent of land	1.0	113.9	114.9
- Depreciation cost of equipment	–	99.3	99.3
- Opportunity cost of equipment	–	79.0	79.0
Total costs per hectare	1 102.9	805.6	1 908.5
Total costs per kilogram			1.58
Total variable costs/kg			0.76
Total fixed costs/kg			0.82
Yield per hectare (kg)			6 819.0
Price of fruit/kg (US$)			0.72
Income per hectare (US$)			4 904.36
Profit per hectare (US$)			2 995.86

Table 12.5 shows the costs and returns of one durian orchard in Phang-Nga province. However, durian orchards have been established in many locations and in many countries. It is unlikely that current prices will stand, but even when current prices are discounted, returns compare favourably with other investments. Thus, it is essential for any prospective investor or grower to budget for their individual circumstances. The time taken until income exceeds costs

depends on location, cultivar and orchard management. However, if good financial records are kept, it is possible to project an itemised production costs figure. If a general industry production costs figure was available, growers could compare their figures to it to judge the efficiency of their own orchard.

13

Post-Harvest Technology

Durian are highly perishable. Therefore, post-harvest technology plays an important role in getting the fruit from the orchard to the consumer, particularly in the case of distant markets. Durian are commercially grown in only a few Asian countries where post-harvest physiology and technology are not yet as well-established as in Australia, the USA and European countries. As a result, only limited information on durian post-harvest physiology and technology is available.

Harvesting Index

Fruit maturity plays an important role in determining the quality of fresh fruit. The taste of the ripe durian aril will be inferior if the fruit is harvested too early or too late. It is very important to harvest durian at optimum maturity. Whether the durian is for fresh consumption or to be processed, whether it is intended for a local or distant market, and what kind of storage facilities are available, are all factors that must be taken into account when determining optimum maturity. There are many ways to determine maturity, but not all of them are reliable.

Calendar

The time taken from first flowering, full bloom or fruit set to maturity depends on the cultivar. Early cultivars such as Kradum Thong, Chanee and Luang take 95 to 105 days to mature from full bloom. Medium cultivars such as Kan Yao, Chompoo Si and Tong Yoi Chat take 105 to 120 days, and late cultivars such as Mon Thong, Kob and E-Nak take 120 to 140 days

(Somsri, 1987; Chaiwongkiat, 1992). However, the calendar is not an entirely reliable guide because commercial maturity largely depends on location, temperature, soil fertility and cultural practices (Chaiwongkiat, 1992; Nanthachai *et al.*, 1994).

Fruit colour

As the fruit approaches maturity, the colour at the base of the spines changes from dark green to light green, or a greyish-green with brown tips. The more mature the fruit, the browner it becomes. The groove at the base of the spines also browns as the fruit matures.

Sound

If the fruit is mature, there is a hollow sound when either the fingertips, a knife or a wooden stick are run through the spines. If the fruit is not yet mature, there is a more solid sound. As the fruit matures and becomes lighter, the aril lifts away from the husk leaving a space which accounts for the hollow sound.

Spines

As previously mentioned, the tips of the spines brown upon maturation. The spines also become more flexible and can be pushed inwards more easily than those of immature fruit. This is also due to the loss in weight.

Odour

The fruit emits its characteristic odour when it has begun to ripen. The method of using the odour to detect maturity is used by growers who want to harvest half or three-quarter ripened fruit for local markets. However, fruit harvested at an early stage of ripening will not be suitable for distant markets or long-term storage.

Harvesting Techniques

Mature fruit can be harvested at any time of the day except when it is raining. Water will cause the fruit to change colour from light green to dark green and it then becomes difficult to judge the maturity of the fruit. The wet ground also makes harvesting in the rain an uncomfortable process.

Because durian flowers bloom at different times, not all fruit on the same branch will mature at the same time. Therefore, fruit maturity must be checked individually before harvesting. Harvesting should be done every day or every other day.

Since durian trees flower from the branches or trunk, fruit is borne inside the canopy. (This differs from mango, mangosteen, rambutan, lychee, longan, etc, which flower from the apical meristems.) It is therefore difficult to harvest durian outside the canopy. Growers must climb into the canopy and harvest mature fruit using a knife or scissors to cut the fruit stalk above the abscission mark (Plates 13.1a and b). Individually harvested fruit are then dropped into a jute sack which acts as a cushion. Two people can harvest 2 000 to 2 500 fruit per day using this method. However, if the tree is very tall, it can be difficult to drop the harvested durian into the jute sack. In this case, a rope is tied to the durian stalk and the fruit is lowered to the ground. If the proper knot is used, the rope will automatically loosen and the fruit is released. One person can harvest 600 to 700 fruit per day using this method. Furthermore, the harvested fruit do not risk being damaged by missing the jute sack and falling to the ground as is possible using the former harvesting method (Anon, 1989).

In the Philippines, Indonesia and Malaysia, most fruit are allowed to fall to the ground (Coronel, 1983; Nanthachai *et al.*, 1994). There are two possible reasons for this: first, the fruit-bearing trees may be very tall, and secondly, people in these countries prefer to eat fully or over-ripe durian.

Post-Harvest Physiology

Weight loss in durian during ripening under ambient temperatures (28°C to 31°C) is approximately 13% to 21% (Sriyook, 1990; Ketsa and Pangkool, 1994). This is high compared with data obtained for tangerines (Ketsa, 1990), mangoes (Pantastico *et al.*, 1984) and sugar apples (Kasiolarn, 1991). However, despite a 21% weight loss in ripened durian over a nine day period at ambient temperatures (30°C and 70% relative humidity (RH)), the fruit is still marketable (Sriyook, 1990). This may be due to the thick and extremely tough husk which shows no sign of shrinkage even after such a huge weight loss.

Most weight loss occurs in the husk (rind) tissue. There is no significant change to the moisture content of the aril in different humidities. Furthermore, the moisture content of the husk steadily decreases, while the moisture content of the aril steadily increases during ripening (Daengkanit, 1995a). There are occasionally arguments between buyers and sellers because of this large change in the net weight of the durian between production site and market. This suggests that there is a lack of information on the natural post-harvest weight loss of durian.

Dehiscence

Dehiscence is associated with ripening and tends to occur when the fruit is fully ripe or over-ripe (Sriyook, 1990). Dehiscence takes place along the suture on the back of the locules, which have an abscission zone (Praditkul, 1987). The blossom end of the fruit is usually first to dehisce, but any point along the suture is vulnerable (Plate 13.2). Dehiscent fruit cannot be kept long, particularly in warm conditions, because the aril is easily infected by micro-organisms.

Dehiscence is greater in low humidity than in high humidity (Ketsa and Pangkool, 1995b). As a result of dehydration, the tissue in the suture region may shrink in different directions from that of adjacent tissue. The tension that develops assists the opening of the suture (Ketsa and Pangkool, 1994). Endogenous ethylene is thought to be a natural stimulant in durian dehiscence (Sriyook *et al.*, 1994), however, this is yet to be proven. Coating or waxing the surface of the fruit to lessen weight loss can reduce dehiscence (Sriyook, 1990).

Respiration and ethylene production

Durian is a climacteric fruit (Tongdee *et al.*, 1988a; Booncherm, 1990). The rise in respiratory activity and ethylene production coincides with ripening (Ketsa and Daengkanit, 1998b). Climacteric respiration increases from 24°C to 33°C, while the ethylene peak reaches a maximum at 27°C and declines rapidly as the temperature increases from 30°C to 33°C (Ketsa and Pangkool, 1995a). This indicates that ethylene production in durian appears to be more sensitive to high temperatures than in climacteric respiration. Similar results have been reported for apple (Lurie and Klein, 1990), avocado (Eaks, 1978) and tomato (Buescher, 1979). The mechanism of high temperature decline in ethylene production of durian may involve inhibition of 1-aninocylopropane-1-carboxylic acid (ACC) to ethylene (Yu *et al.*, 1980; Atta-Aly, 1992) as the ACC oxidase, responsible for the conversion of ACC to ethylene, is heat-sensitive (Chan, 1986a, b).

According to market practice, fruit are most commonly harvested at 85% maturity. In cultivars such as Chanee, this is about 110 days after fruit set. The fruit will show a climacteric peak within four to five days after harvesting. Fruit harvested at a stage later than 85% maturity exhibit the climacteric rise almost without a preclimacteric lag period. The relatively high rate of respiration and ethylene production in fruit harvested at this stage indicates that ripening is well underway at harvest, and the fruit is ripe enough for consumption on the second or third day after harvest (Tongdee *et al.*, 1988b).

When the aril and husk of Chanee durian are separated, respiration and ethylene production of the husk are much higher than that of the aril. The climacteric pattern of respiration and ethylene production is also inhibited in the husk but not in the aril. The softening of the separated aril is slower than that of the aril in intact fruit. This indicates that the ripening of durian aril requires ethylene from the surrounding husk (Booncherm, 1990).

ACC synthase and ACC content in both Chanee and Mon Thong durian increases as the fruit ripens. ACC synthase then declines in both the aril and the husk, while ACC content increases to the highest level on the final day of storage. ACC synthase in the husk is higher than that in the aril, while ACC content is higher in the aril. ACC oxidase activity increases as the fruit ripens and decreases towards the end of storage. This suggests that ACC oxidase activity in the husk is greater than that in the aril (Chaiprasart, 1993), resulting in more ACC converted to ethylene, which in turn triggers the ripening of the aril.

Carbohydrates

Starch content decreases concomitantly with a rapid increase in soluble solids and total sugars as ripening advances (Ketsa and Daengkanit, 1998b). This decrease in starch, accompanied by an increase in soluble solids and total sugars, is typical of post-harvest carbohydrate changes in climacteric fruit (Biale, 1960). The total sugar content of the aril of ripened durian is rather high at 21% to 26% (Sutthaphan, 1993; Ketsa and Pangkool, 1994, 1995a). This makes durian very sweet, with a rich taste unlike any other fruit. Reducing sugars (glucose and fructose) rapidly increase with advanced ripening. The aril of ripening durian contains 2% to 5% reducing sugars depending on the cultivar. Chanee has more reducing sugars than

Mon Thong. However, Mon Thong aril contains more non-reducing sugar (sucrose) than that of Chanee with non-reducing sugars contributing 70% to 80% of the total sugars. Sucrose, therefore, has a more important role in sweetening the aril of ripening durian (Sutthaphan, 1993).

Pigmentation

Durian aril changes from light to dark yellow as it ripens, depending on the cultivar (Ketsa and Pangkool, 1994). This may be due to an increase in β-carotene synthesis in the aril during ripening (Ketsa and Pangkool, 1995a). The pulp of ripening Chanee durian contains two to three times more β-carotene than that of Mon Thong durian (Sutthaphan, 1993). This explains why the aril of Chanee durian is a darker yellow than that of Mon Thong.

Softening

The firmness of the aril rapidly decreases during ripening (Ketsa and Pangkool, 1994; 1995a, b) and is accompanied by a rapid increase in water soluble pectin (Ketsa and Daengkanit, 1998b). Polygalacturonase and pectinesterase activities increase as the aril softens in the ripening durian. In unripe fruit polygalacturonase activity is very low and, as the durian ripens, increases more dramatically than the pectinesterase activity which is already quite high in the unripe fruit. β-galactosidase activity is also very high in the unripe aril and the levels slightly increase as the fruit ripens. Cellulase activity is equally high in the unripe aril but remains constant as the aril softens. This suggests that the integrated action of polygalacturonase and pectinesterase is required for the durian aril to ripen, and that both β-galactosidase and cellulase are not involved in this process (Ketsa and Daengkanit, 1998a; 1998b; 1999).

Odour

The distinctive durian aroma, which indicates the ripeness of the fruit, can be detected one day before the respiratory peak (Tongdee *et al.*, 1988a). The odour has two distinguishable components: one strong and onion-like, the other more delicate and fruit-like. If the fruit is kept in an enclosed space, a third component, better described as a 'stink', can then be detected (Martin, 1980). Thiols and thioethers are responsible for the onion-like odour and esters for the fruit-like odour. The chief thiols in order of importance are propanethiol, ethanethiol and methanethiol. The ester compounds included 17 components. One ester predominated over all thiols and had a pleasant fruity smell; this was identified as ethyl 2-methylbutyrate (Baldry *et al.*, 1972).

Post-Harvest Handling

Sorting

At harvest, the fruit show considerable variation in quality. Therefore, growers and traders must cull their fruit in order to get suitable returns commensurate with quality. Quality variation is associated with abnormal shape and size, defects such as broken stalks (which frequently occur at abscission), loose stalks, broken spines and infestation with mealybugs, scale insects or sooty mould. The fruit must be sorted manually. Culled fruit are sold at low prices for local fresh consumption or for processing.

Grading

At present, there is no validated grading or standards set for durian in durian-growing ASEAN countries. However, growers, traders and particularly exporters realise how important grading is. The culled durian are manually graded according to their size. A durian grading machine has been developed (Jarimopas *et al.*, 1992), but it is not used commercially. After grading, durian are sized for uniformity.

In Thailand, growers and traders voluntarily grade durian according to the requirements of the market (Ratanachinakorn and Nanthachai, 1994). For example, Europe, Canada and the USA prefer 2.5 to 3.5 kg fruit packed 10 to 12 kg per carton. Hong Kong prefers 2 to 4 kg fruit packed in 35 to 50 kg bamboo baskets. Malaysia and Singapore prefer 2 to 5 kg fruit packed either in bamboo baskets, wooden crates, or packed loose on a truck tray.

The Thai Department of Agriculture (Hiranpradit *et al.*, 1992) has proposed criteria for informal quality standards of export durian for the three main commercial cultivars in Thailand: Chanee (Table 13.1), Mon Thong (Table 13.2) and Kan Yao (Table 13.3). Draft classifications according to shape are shown for Chanee (Figure 13.3), Mon Thong (Figure 13.4) and Kan Yao (Figure 13.5).

Table 13.1:* *Quality standardisation of Chanee durian in Thailand (Hiranpradit et al., 1992)

	Class		
Items	*Extra*	*I*	*II*
External fruit shapes	*1, 2, 3+*	*3*	*1, 2, 3+, 3, 3–*
Fruit weight (kg)	*2.0–3.5*	*2.0–3.0*	*3.5–4.5, >3.0, 2.5–3.0, 1.0–2.0, 1.0–2.0*
Fruit health	*– free from disease and insects – slight superficial defects, not markedly visible*	*– free from disease and insects – slight defects, not affecting internal quality*	*– free from disease and insects – slight defects, not affecting internal quality*
Internal quality	*– no wet core – no browning*	*– no wet core – no browning*	*– no wet core – no browning*
Flesh colour	*– yellow*	*– yellow*	*– yellow*

Remarks
For all classes : good odour and flavour with fine texture
Extra class : 5% by number or weight of durian not satisfying the requirement of the class but meet class I
Class I : 10% by number or weight of durian not satisfying the requirement of the class but meet class II

Table 13.2: Quality standardisation of Mon Thong durian in Thailand (Hiranpradit et al., 1992)

	Class		
Items	Extra	I	II
External fruit shapes	Mon 3/2	Mon 3/2	3/2
Fruit weight (kg)	2.0–4.5	4.5–6.0	2.0–4.5
Fruit health	– free from disease and insects – slight superficial defects, not markedly visible	– free from disease and insects – slight defects, not affecting internal quality	– free from disease and insects – slight defects, not affecting internal quality
Internal quality	– no wet core – no browning	– no wet core – no browning	– no wet core – no browning
Flesh colour	– yellow	– yellow	– yellow

Remarks
For all classes : good odour and flavour with fine texture
Extra class : 5% by number or weight of durian not satisfying the requirement of the class but meet class I
Class I : 10% by number or weight of durian not satisfying the requirement of the class but meet class II

Table 13.3: Quality standardisation of Kan Yao durian in Thailand (Hiranpradit et al., 1992)

	Class		
Items	Extra	I	II
External fruit shapes	Lychee, 3/2	Lychee, Round, 3/2	2/3
Fruit weight (kg)	3.01–4.5	2.0–3.0, 2.5–4.5, 2.01–3.00	1.5–2.0
Fruit health	– free from disease and insects – slight superficial defects, not markedly visible	– free from disease and insects – slight defects, not affecting internal quality	– free from disease and insects – slight defects, not affecting internal quality
Internal quality	– no wet core – no browning	– no wet core – no browning	– no wet core – no browning
Flesh colour	– yellow	– yellow	– yellow

Remarks
For all classes : good odour and flavour with fine texture
Extra class : 5% by number or weight of durian not satisfying the requirement of the class but meet class I
Class I : 10% by number or weight of durian not satisfying the requirement of the class but meet class II

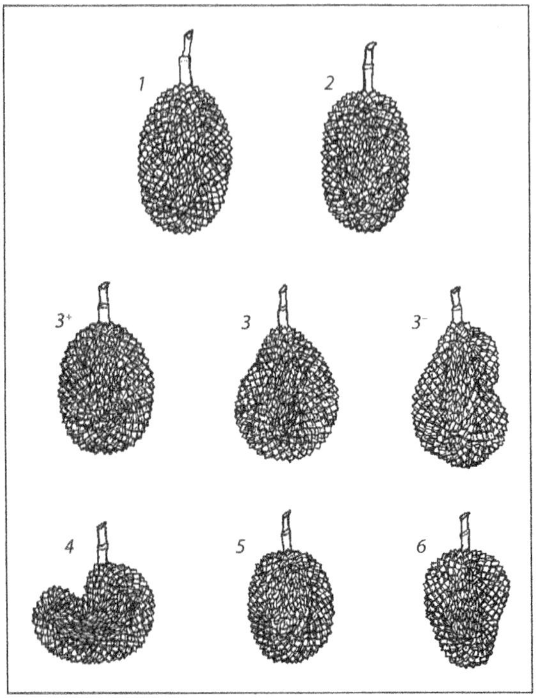

Figure 13.3 Classification of Thai Chanee durian fruit shape is as number 1, 2, 3, 4, 5 or 6 (Hiranpradit et al., 1992)

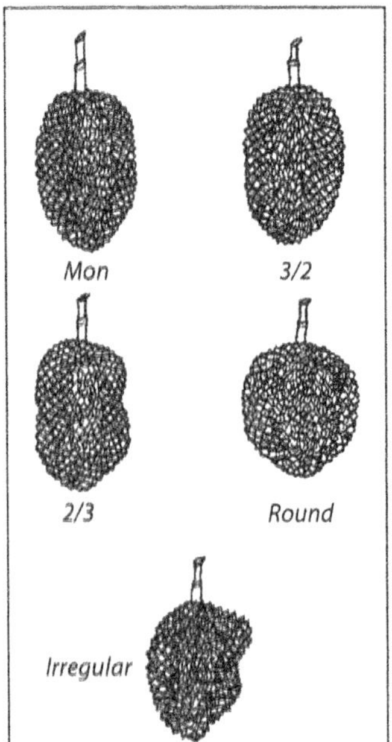

Figure 13.4 Classification of Thai Mon Thong durian fruit shape is as number 3/2, 2/3, Mon, Round or Irregular (Hiranpradit et al., 1992)

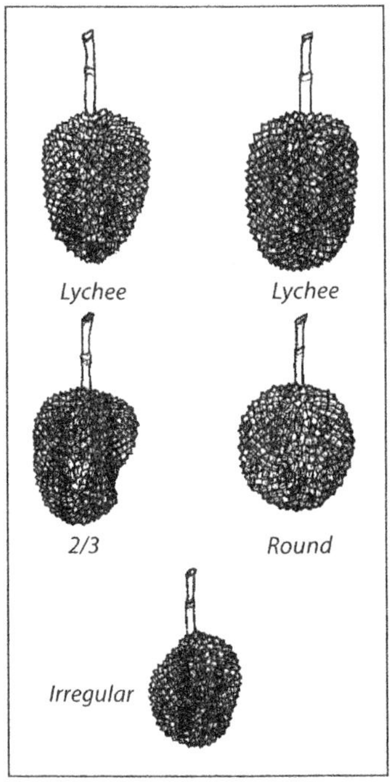

Figure 13.5 Classification of Thai Kan Yao durian fruit shape is as number 2/3, Lychee, Round or Irregular (Hiranpradit et al., 1992)

Storage

Durian fruit harvested at approximately 85% maturity can be kept at room temperature (29°C to 31°C) for three to six days. This depends on the cultivar; early cultivars such as Chanee can be kept for three to four days, while late cultivars such as Mon Thong can be kept for five to six days. Once the fruit is fully ripe, it can be kept for one to two more days but after this the aril will be too soft with a strong sulphury odour.

Mathur and Srivastava (1953) reported that durian can be stored at 4°C to 5°C and at 85% to 90% RH for 30 to 35 days. Durian stored at this temperature appear to live a considerable time without chill damage. However, this may be because only the aril of ripe durian was stored. Praditdoung (1986) obtained a similar result. She found that the aril of the Chanee cultivar, when kept in polystyrene trays and wrapped in low density polythene film, could be stored at 4 ± 1°C and 80 ± 5% RH for 30 days and still remain suitable for eating.

When the cultivar Mon Thong was stored in an atmosphere containing 5% to 7.5% oxygen at 2°C for one week, fruit ripening was inhibited. Under such conditions, respiration and ethylene production remained at the low pre-climacteric level. When the fruit was brought back to room temperature, there was a sudden increase in respiratory activities and ripening developed uniformly. The quality of a fruit ripened in this way was slightly affected but remained acceptable. Lowering the oxygen level to 2% has been observed to injure the fruit (Tongdee and Suwanagul, 1988). Khunprom (1995) found that when Mon Thong durian were stored in an atmosphere containing 5% to 7.5% oxygen at 15°C for three weeks, they were still edible once they were allowed to continue ripening at room temperature.

Mohamed (1990) reported that dark patches (a sign of chill damage) appeared on the husk of the D 8 cultivar by the third week of storage at 8°C. Siripanich (1993) found that when the Chanee cultivar was stored at 5°C, 10°C and 12.5°C for three to ten days, it showed signs of chill damage but when kept at 15°C no damage was seen. In addition, all mature fruit of the Chanee and Mon Thong cultivars showed signs of chill damage when stored at 13°C. They ripened abnormally, contained a low amount of soluble solids and had yellow, brown and grey patches on their husks. The optimum temperature for storing durian is 15°C and they can be stored at this temperature for two weeks without developing chill damage (Siripanich, 1993).

Waxing

Durian fruit lose a great deal of weight when kept at room temperature. If this weight loss cannot be controlled, growers, wholesalers, retailers and exporters risk losing significant amounts of money. Surface coating or waxing has been observed to considerably reduce weight loss in durian fruit. Coating materials such as Sta Fresh®, Semperfresh®, Apple Wax® and Storage Wax® have been shown to reduce weight loss in durian fruit by 7% to 14%. Semperfresh® (at a concentration of 1%) was the most effective; it reduced weight loss by approximately 14% during nine days storage at 29°C (Sriyook *et al.*, 1994).

Waxing restricts gas exchange through the husk resulting in high carbon dioxide levels and lower internal oxygen and ethylene concentrations compared with non-waxed durian fruit, delaying ripening by creating a modified internal atmosphere (Tongdee *et al.*, 1990). If a lower temperature of 15°C is used, waxing is of no additional benefit (Chaiprasart, 1993; Sutthaphan, 1992). Waxing also reduces fruit dehiscence (Tongdee *et al.*, 1990; Sriyook *et al.*, 1994); due either to the reduction in weight loss or the inhibition of ethylene production, depending on whether water stress (causing weight loss) or ethylene is the primary cause of fruit dehiscence.

Fruit stalk preservation

Many growers and retailers preserve the durian stalk by wrapping the stalk with banana leaves (Plate 13.6) or newspaper to protect the stalk from breaking or being otherwise damaged (Kosiyachinda *et al.*, 1995). Another reason for this is to preserve the freshness of the stalk and prevent it losing too much water and becoming shrivelled. Some consumers check how fresh the durian is by looking at the stalk. If the stalk does not appear fresh, it would also appear that the durian is not freshly harvested. Thus, preserving the freshness of the stalk is important in attracting consumers.

Ripening

As previously stated, it has been found that durian harvested at 85% maturity will ripen within three to six days. However, in comparison with other fruits, few comprehensive experiments regarding the effects of post-harvest humidity, temperature and/or biochemical changes have been conducted on durian (Nanthachai, 1994). The quality of ripened durian depends on several factors, of which the ripening regime (including temperature and humidity) of the fruit is of particular importance.

Temperatures between 30°C and 33°C have been found to be best for ripening the fruit of the Chanee cultivar. When the temperature is lowered, the quality declines (Ketsa and Pangkool, 1995a). There are both advantages and disadvantages to ripening durian under low or high RH. Durian ripened under low RH are very easily dehusked but the weight loss is high; at high RH dehusking is very difficult but weight loss is low. In addition, when durian are ripened under low RH, most of the greenish unripe colour on the rind disappears (Ketsa and Pangkool, 1995b) and it is therefore obvious to consumers that the fruit has ripened. On the other hand, the rind of durian fruit ripened under high RH remains green and consumers are less likely to purchase these fruit because it appears they have not yet ripened (Ketsa and Pangkool, 1994).

Durian are frequently harvested when they are not fully mature, usually at 75% to 85% maturity. As durian is a climacteric fruit (Tongdee *et al.*, 1988a; Booncherm, 1990), ethylene is necessary to induce ripening. Many growers induce ripening using traditional methods before selling the fruit. They may cover durian fruit with jute sacks or old blankets, or blanket the fruit between fresh tree leaves. These conditions can increase the temperature and ethylene concentrations surrounding the durian, which in turn stimulates ripening. Some growers spray durian with ethyl alcohol, as under high temperatures ethyl alcohol breaks down into ethylene (Abeles *et al.*, 1992).

Ripening can also be induced by dipping the durian in a solution of ethephon (500 to 3 000 ppm) before transporting the fruit to market. Some growers try to dip only the stalk in the ethephon solution in order to preserve the solution but this is less effective than dipping the whole fruit. The treated durian ripen and are of good quality for eating upon arrival at market.

The use of ethylene gas to induce ripening in durian is not yet commercially practised. There are many reasons for this: 1) ethylene gas is not readily available; 2) it is still very expensive; 3) operating costs are also very high; and 4) the application of ethylene gas is not a simple procedure. The gas must be introduced into an airtight ripening room in which the temperature and humidity are controlled. One should also take into consideration other problems that may arise in using ethylene gas to ripen durian. Since durian has higher rates of respiration and transpiration than other ripening fruit, the subsequent concentration of carbon dioxide (which prevents ripening) and increase in RH (which induces ripening) in the ripening room will both inhibit ripening and cause fruit rot. Therefore, ripening rooms should not be over-crowded and should be well-ventilated to prevent the build-up of both carbon dioxide and ethylene gas. Too much ethylene can turn the durian rind yellow which is not acceptable to the consumer (Siripanich *et al.*, 1990). Ethylene treatment tends to have a more pronounced effect on aril colour and firmness, than the soluble solids, sugars and starch contents of the aril (Ketsa and Pangkool, 1996).

Packaging

Durian is considered to be one of the largest and heaviest fruits; its husk is spiny and rigid. Therefore, durian packaging must be very strong in order to get the fruit to their destination. Generally, two types of packaging are used for durian; one is a harvesting container and the other is for shipping purposes. Most growers do not use harvesting containers. After the harvest, durian are usually laid on the ground on jute sacks or plastic sheets beneath the fruit trees (Plate 13.7a). They will then be loaded on a cart which takes them to a nearby packing house (Plate 13.7b). Only a few growers are willing to accept advanced technology by using harvesting containers or bulk bins made of wood (Plate 13.8a) or metal (Plate 13.8b) to make production more efficient.

Many types of shipping packages or containers are used for durian depending on whether the market is local or distant. Bamboo baskets (Plate 13.9) are widely used for the local market throughout ASEAN countries because they are cheap and readily available. One bamboo basket can hold 30 to 50 kg of durian. However, durian fruit in bamboo baskets are easily damaged as the baskets are not strong enough to withstand much weight when they are stacked upon each other. Despite this, some exporters still use bamboo baskets to ship durian to distant markets in Singapore and Hong Kong.

Wooden crates are also used for shipping durian to local markets but they are not as popular as bamboo baskets. They are much heavier, more expensive, less capacious and are less readily available.

Corrugated boxes are commonly used for distant markets (Plate 13.10). They are standard packaging and accepted worldwide. Each box can hold four to six fruit depending on the size of the fruit. Cardboard may be used to partition the fruit in order to prevent them moving during transportation.

In addition to the containers mentioned above, durian are often shipped in bulk from orchards to local wholesale markets (Plate 13.11) without containers. Wholesalers and retailers who buy durian directly from growers usually prefer to transport durian in this way. More durian can be shipped and there are no packaging costs. However, loading and unloading the individual durian fruit is much more time-consuming.

Transportation

Once harvesting has finished, fruit must be transported to the markets as quickly as possible in order to avoid over-ripening, fruit rot and weight loss. The fruit are usually transported to local markets on trucks with no way of controlling temperature or RH. Fruit are either loaded individually on to the truck or packed into bamboo baskets or wooden crates. In Thailand, fruit are preferably transported individually in bulk (Plate 13.11).

Durian may be exported by truck, ship or air depending on the market. Durian from Thailand are transported by truck to Malaysia and Singapore. Tarpaulins are used as a cover to protect the fruit from rain and direct sunlight during transit.

Durian are exported to Hong Kong, Japan, Europe, Canada and the USA by air. Bamboo baskets are the preferred packaging for the Hong Kong market whereas Japan, Europe, Canada and the USA prefer cardboard cartons (Kosittrakun and Nanthachai, 1994). The fruit must reach its destination before it ripens or the smell will become a serious problem for aircraft passengers. Basil leaves (Vejjajiva, 1986) and activated carbon have been reported to reduce the smell of durian (Srikumlaithong and Svaditat, 1990) but neither of these is commonly used commercially. Thai exporters generally overcome the odour problem by harvesting durian at exactly the right time so that the fruit will become ripe on the same day it reaches the market or a few days after arrival.

Sea transportation is cheap and readily available but it can be used only for shipping durian to Hong Kong at present. With the more distant markets, durian will ripen during transit. Durian cannot be transported under refrigeration because of the danger of chill damage (Siripanich, 1993). Occasionally, durian have been transported by sea in 'dry' containers with no method of temperature or humidity control. The only way to avoid abnormal ripening, fruit rot and dehiscence in this situation is to leave the back door of the container open to allow air to circulate (Kosittrakun and Nanthachai, 1994).

14

Durian Products

Many tonnes of fresh durian arrive on the market every year from producing countries such as Thailand, Indonesia and Malaysia. The price of this fruit fluctuates annually depending on the supply. If production is plentiful, then the price of durian will be low and vice versa. This encourages people to seek new outlets for surplus produce by developing products from the excess of fresh fruit. As these new durian products become increasingly popular in Thailand, their market price also becomes more attractive for the producer.

In addition to creating a new outlet for surplus durian, other motivations for developing durian products are to:

- develop value-added products;
- extend the shelf-life of durian (durian products keep longer than fresh fruit);
- extend the marketing period;
- solve the odour problem (unlike fresh durian, durian products can be shipped by air without becoming odorous);
- reduce the cost of shipment (durian products are less heavy and bulky than fresh durian);
- improve the durian cultivation system (ensuring yield uniformity and quality of both durian products and fresh fruit);
- encourage the sale and promotion of durian within both the local and export markets.

Durian aril can be processed into several products as described below. These products either preserve the fresh aril in its original state or change it into an entirely new form. More durian products have been developed in recent years, possibly because new ways of eating the fruit have been so well received.

Minimally Processed Durian

The edibility of aril from minimally processed durian is the same as that of fresh durian. This minimally processed, ready-to-eat durian is easier for people who have difficulty dehusking the fresh fruit. (Plate 14.1 shows what the product looks like when displayed in the market.) The durian needed for minimal processing should be half-ripe before being dehusked. If it is still unripe the minimally processed aril will not ripen normally and if over-ripe, it will be too soft and become watery.

The durian aril with seeds intact is removed from the locules and placed in a foam tray before being wrapped with polyvinyl chloride (PVC) film to prevent condensation inside the package. Paper towel is sometimes laid on the foam tray beneath the aril to reduce humidity within the package and prevent the aril becoming watery. Caution should be taken in the minimal processing of durian to ensure that conditions remain hygeinic to prevent microbial contamination which causes rapid decay, particularly at room temperature.

Aril wrapped in PVC can be kept longer at lower temperatures and, if stored below 10°C, will not suffer the same chill damage sustained when the whole durian is frozen.

Frozen Durian

Freezing is considered the best method for preserving the appearance of the fresh aril during long-term storage. Durian can be frozen by either contact freezing or individual quick freezing (IQF) as whole fruit or seed-free aril. Freezing whole durian fruit is less preferable because of the large freezing area and high energy input required. Durian frozen by IQF can be kept for

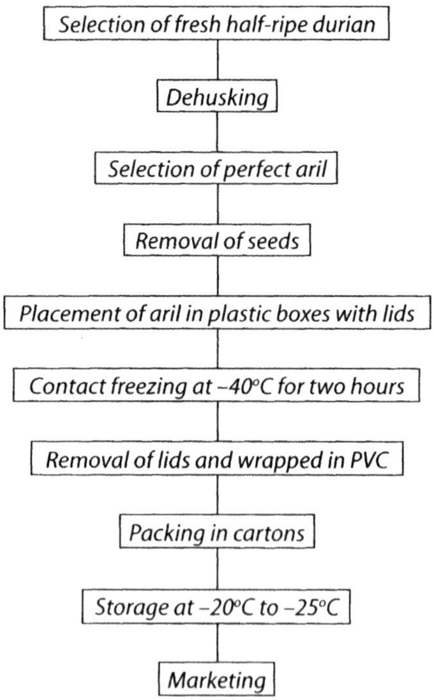

Figure 14.2 *Contact freezing process for preserving durian*

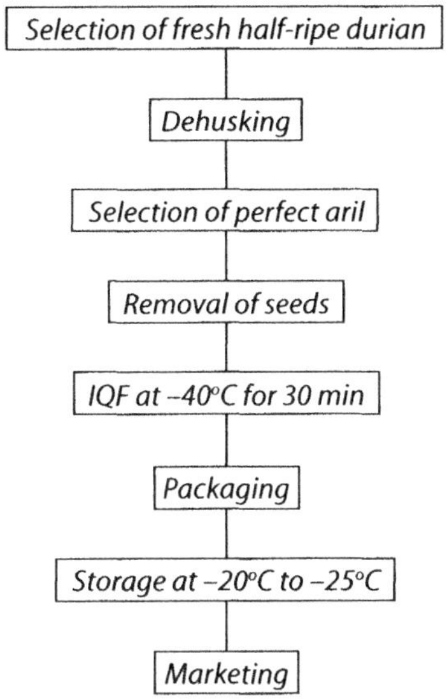

Figure 14.3 *Individual quick freezing (IQF) process for preserving durian*

a year whereas contact frozen durian can be kept for only six months, turning a dark colour if it is kept any longer. The step-by-step process of contact freezing (Figure 14.2) and IQF (Figure 14.3) is similar, from the aril preparation through to the end product, varying only in the method of freezing. Frozen durian is becoming more popular for export markets because the strong smell is eliminated and it can be shipped by sea, a less costly alternative to air transport.

To serve frozen durian, the aril can be thawed in a microwave or oven for three to four minutes at a moderate heat. (Thawed durian aril looks much like fresh durian.)

Durian Powder

Durian aril can be processed into a powder which is used as an essence in coconut milk, icecream, cakes, cookies, biscuits, candy, beverages and other products requiring durian flavour. Durian powder is made from unripe but fully mature fruit. Such fruit should be processed into powder within 48 hours of harvesting as ripe fruit is unsuitable for manufacturing this product. To obtain 1 kg of durian powder, approximately 3 kg of fresh durian is necessary (Haruthaitanasan, 1987). The flow chart in Figure 14.4 shows the processing of durian into powder.

Durian powder should be stored in a polypropylene bag within a tightly sealed container and kept in a cool place with low humidity. Packed in this way, durian powder will keep for a year without its colour or viscosity changing.

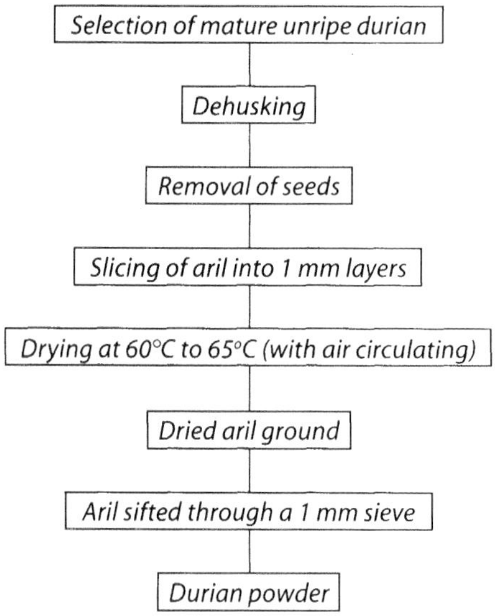

Figure 14.4 *Processing of durian to produce powder*

Durian Paste

In Thailand, durian paste has been made for many centuries by growers. Durian paste is made from surplus over-ripe or dropped fruit. Over-ripe durian is preferable for paste because it is easier to dehusk and stir during mixing. Durian paste can also be made from frozen aril but will take longer to heat because it has a higher moisture content than fresh aril. Durian paste can be made from the fruit of individual or mixed cultivars.

Over-ripe durian aril is poured into an open pan, along with 12% to 20% sugar by weight, and heated over a wood, charcoal or gas stove for approximately three hours (depending on the quantity). The heat should be moderate to avoid burning and the paste continuously stirred with a paddle to mix the aril and to remove moisture (Figure 14.5). To check whether the paste is ready, a small amount should be picked up by hand during stirring. If it is not sticky to the touch, then stirring may stop and the paste left to cool. If it is still sticky, stirring must continue.

The finished product can be wrapped in plastic film and packaged in a foam tray depending on its weight and which market it is intended for (although the aroma, colour and flavour of durian paste is better preserved if kept in laminated aluminium foil). Durian paste is usually sold wrapped in plastic film and comes in different sized packages ranging from 200 to 500 g (Plate 14.6). Durian paste can be kept at room temperature for four months or frozen at -20°C for one year without a reduction in edibility (Paveenkarn, 1988).

Durian Chips

Durian chips have recently been developed from immature or unripe durian fruit (Figure 14.7). They have a growing base of consumers who believe they taste better than potato chips. Durian chips can even be made from dropped fruit or insect infested unripe

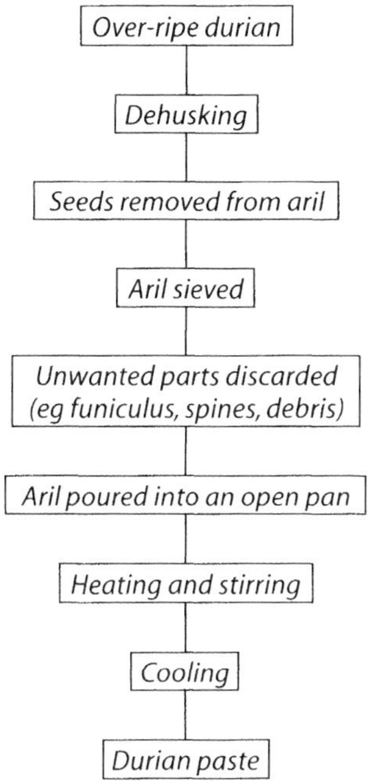

Figure 14.5 *Processing of durian to produce durian paste*

fruit. In fact, the most attractive chip is produced from the light yellow aril of unripe fruit. The de-seeded aril is sliced into thin layers approximately 1 mm thick and deep fried at a moderate heat for 8 to 10 minutes. These chips are then removed from the oil and left to dry on paper tissue to absorb the extra oil then served either with or without salt.

Durian chips are graded into different categories depending on their size and packed in cans, plastic bags or plastic boxes (Plate 14.8). Chips will keep longer if stored in a place where warm air circulates to reduce moisture content.

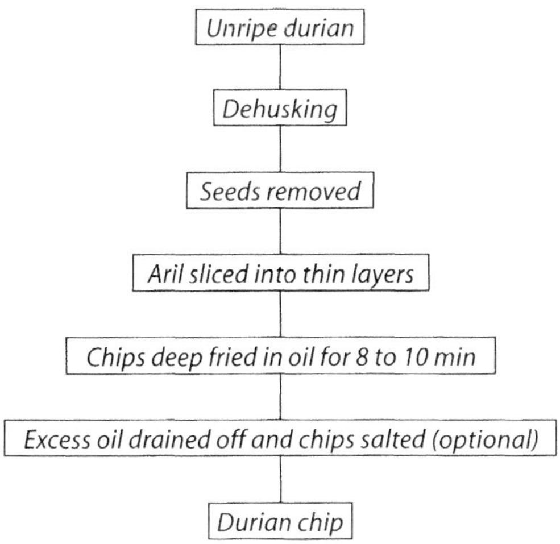

Figure 14.7 *Processing of durian to produce durian chips*

Other Products

In addition to the durian products mentioned in this chapter, other durian products, both savoury and sweet are manufactured and many of these are popular in cafes, hotels and restaurants (Thailand Institute of Science and Technology, 1997). The Chantaburi Horticultural Research Centre in Thailand (1996) has published 32 recipes using durian aril, and durian products are well-known and accepted in such dishes as salad, tempura, chowder, cake, icecream, cookies, candy and biscuits. In fact, durian products can be made from fruit at any stage of development. Even the mature seeds can be roasted and eaten as a delicious snack.

15

Marketing

Durian production is confined almost exclusively to tropical South-East Asia. Only Thailand produces a large enough quantity of durian to supply both local and export markets. Other countries such as Malaysia, Indonesia and the Philippines do not produce enough for local market demands and therefore must import fruit, mainly from Thailand.

The availability of durian is dictated by the short harvest season which is typically only two to three months long, although small quantities are harvested during other months of the year. In addition, durian has a short shelf-life and cannot be stored for long periods at low temperatures. These factors mean that the fresh fruit is available on the market for only six months (late March to September). However, durian products including frozen durian and durian paste are available all year round.

Thailand

Local market

Durian production in Thailand increased every year between 1988 to 1994. Total consumption has also increased (Table 15.1). However, most of the produce was locally consumed and less than 5% was exported in the form of fresh durian and durian products.

Durian is seasonal, like other tropical fruits such as lychee, longan, mango, mangosteen and rambutan. The harvesting of Thai durian begins in late March and continues until September, with a peak harvest period between May and August. This is due to the diversity of cultivars and growing locations in Thailand which considerably extend the harvesting season. Fruit from the eastern part of Thailand is harvested and marketed from late March to June while in the south, the season is from July to September (Nonthachai, 1994).

Thailand's considerable durian production results in a variety of marketing opportunities, as can be seen in Figure 15.1.

Table 15.1: Production, local consumption and export of durian in Thailand (Department of Custom Service and Agricultural Extension, Thailand)

Year	Production (tonnes)	Local consumption (tonnes)	Export (tonnes)
1988	444 415	436 588 (98.2%)	8 034 (1.8%)
1989	486 644	475 392 (97.7%)	11 072 (2.3%)
1990	465 315	447 357 (96.1%)	17 909 (3.7%)
1991	539 190	526 870 (97.7%)	12 354 (2.3%)
1992	711 371	689 517 (96.9%)	21 850 (3.1%)
1993	757 583	731 734 (96.6%)	25 849 (3.4%)
1994	767 412	731 148 (95.3%)	36 264 (4.7%)

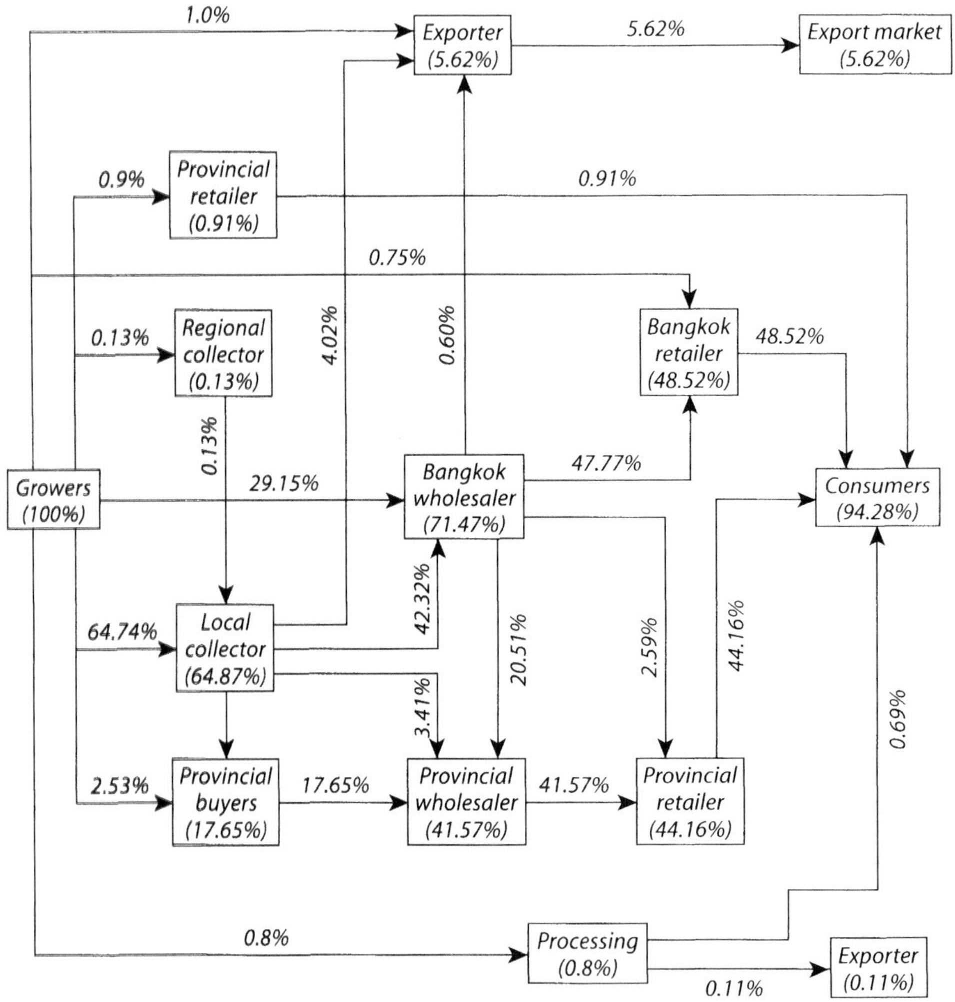

Figure 15.1: *The system for marketing durian in Thailand. Figures in brackets are the total percentages received annually before being distributed (Brunei Darussalam Department of Agriculture, 1988)*

Figure 15.1 indicates that local collectors play an important role in purchasing fruit from growers. Most durian growers sell their fruit to local collectors (about 64.74%), while 29.15% is sold directly to wholesalers in Bangkok, 2.53% to provincial buyers, 1.0% to exporters, 0.9% to provincial retailers, 0.8% to processors, 0.75% to Bangkok retailers, and 0.13% to regional collectors. Wholesalers in Bangkok purchase about 71.74% of total durian fruit production and redistribute this to provincial consumers (20.53%) and retailers in Bangkok (47.77%).

While some fruit for export is purchased directly from reliable growers who produce good quality durian, some exporters will also buy fruit from the personal orchards of local collectors. Acting as both seller and collector, local collectors play a crucial role for both wholesalers and exporters in Bangkok.

With many distribution networks and a big demand for durian from export markets, there is often competition in securing fruit from the orchards. Growers may sell their produce in a variety of ways, as follows :

Contracted orchard for entire crop

Local collectors or provincial wholesalers with enough capital will make contracts with growers and negotiate on the price. Usually this contract price will be signed and payment made when the durian is at 45% to 50% maturity. Advance purchasing of fruit may be done of the whole season's crop, of one harvest only, or on a crop-by-crop basis. The sale of durian in this way reduces the risk of price fluctuation and also reduces marketing problems for growers. However, growers must have some experience in estimating prices close to the market value, or risk losing their income from selling at too low a rate. The total payment will be calculated from the actual weight of the durian at harvest time.

Contracted orchard for partial crop

In this case, collectors or wholesalers will pay growers only once throughout the harvesting season. The sale of durian in this fashion is to the advantage of collectors and wholesalers who are not forced to make an advance payment to growers. Collectors and wholesalers may choose to purchase only the large, high quality durian from the first crop, and not return. If the actual price of durian in the market is lower than the contracted price, they may never purchase the fruit at all.

Free sale

Growers can sell their fruit to anyone if the price is satisfactory. The quantity sold depends on the type of buyer. For example, regional collectors and Bangkok wholesalers will have more purchasing power than local collectors and provincial retailers.

Sale at the market

Growers may take their own fruit by truck or pick-up vehicle to wholesale markets near their orchards. Prices will fluctuate according to negotiations, and also according to market demand and the availability of the fruit.

Local market problems

Although the system of marketing durian in Thailand is progressively developing, some difficulties still remain. These problems are as follows :

Quality

The quality of durian greatly affects sales and therefore the price. Poor quality may be due to physical defects, immaturity, over-ripening and the like. Unfortunately, consumers will not know if the durian they buy is inferior until it is taken home and dehusked. If they find it to be substandard, they may not buy fruit from that particular shop again. It is therefore essential that growers and retailers are reliable.

Transportation

Thailand's durian harvesting season usually coincides with the rainy season and most orchards are located in remote areas with bad road conditions. These two factors cause problems for growers who want to bring their fruit to the local wholesale markets in good condition. Post-harvest losses also occur in transit if packaging and loading are inadequate (Siripanich and Jarimopas, 1993). Most durian delivered to the wholesale markets are not sorted or graded at the production site. As a result, wholesalers must waste a great deal of time sorting and grading fruit upon arrival at the markets.

Price

Durian has a short harvesting season and plenty of fruit is available simultaneously, particularly at the peak of the season. This surplus serves to reduce the price in the markets. In addition, as mature durian ripen quickly (on or off the tree), growers must sell their produce without delay, sometimes regardless of price, or risk the fruit over-ripening and becoming unsuitable for the fresh markets.

Lack of marketing information and facilities

The wholesale market in Bangkok is the biggest and most important market for Thai-produced durian. However, news of its daily prices is not always accessible to the growers and this can result in fruit being sold unnecessarily cheaply. Unfortunately, because the local wholesale markets are not as good as the Bangkok wholesale markets (due to lack of space, sanitation and banking facilities) growers must usually put up with this problem.

Export market

Thailand exports about 5% of its total durian production (Table 15.1). Countries which import this fruit from Thailand are located mainly in Asia (Table 15.2), specifically Hong Kong, Malaysia and Taiwan.

Durian accounts for about 38% of the total fruits exported from Thailand annually. In fact, in the past, the export value of durian ranked second only to that of longan, which it superceded in 1995 (Table 15.3).

***Table 15.2:* Importing countries, quantity and value of durian fruit exported from Thailand (Department of Customs Service, Ministry of Finance, Thailand)**

	1995		1996	
Country	tonnes	US$ million	tonnes	US$ million
Australia	83	0.17	18	0.04
Brunei Darussalam	53	0.12	26	0.04
Canada	33.1	0.05	476	0.68
France	15	0.04	2	0.01
Hong Kong	15 369	9.82	25 368	15.62
Indonesia	278	0.27	491	0.50
Japan	94	0.21	58	0.07
Malaysia	2 343	1.42	2 932	1.86
Philippines	55	0.11	–	–
Taiwan	27 066	26.36	33 304	27.93
UK	2	0.01	4	0.03
USA	267	0.37	339	0.55
Others	249	0.24	358	0.267

***Table 15.3:* Export value of Thailand's fruits (Department of Customs Service, Ministry of Finance, Thailand)**

	Value (US$ million)			
Fruit	1993	1994	1995	1996
Banana	0.66	1.09	0.99	1.26
Citrus	3.53	3.58	3.91	5.12
Durian	15.18	20.90	40.16	59.35
Grape	0.03	0.03	0.02	0.002
Guava	0.02	0.03	0.14	0.12
Longan	15.92	30.26	35.28	51.96
Lychee	2.91	1.46	4.74	13.46
Mango	1.04	1.94	1.69	4.80
Mangosteen	1.25	1.06	2.63	3.86
Papaya	0.06	0.09	0.001	0.01
Pineapple	0.10	0.16	0.28	0.58
Rambutan	0.35	0.62	0.82	1.56
Strawberry	0.34	0.33	0.13	0.002
Sugar apple	0.04	0.60	0.15	0.03
Others	14.78	16.81	23.01	13.97
Total	56.21	78.96	113.95	156.084

Malaysia is Thailand's main competitor for the Singaporean durian market, with Malaysia supplying more than 99% of Singapore's total imported durian. Thailand, however, has the greatest market share of fruit sold in Hong Kong, Japan, Brunei Darussalam, Taiwan and the USA and is increasing the total quantity of fruit it exports each year (Table 15.1). The sales of durian to Japan and the USA is interesting because both these countries have a lot of purchasing power which makes term payment less complicated.

In addition to exporting fresh durian, Thailand also exports durian paste and frozen durian, with exports of the latter increasing every year (Table 15.4). Australia, Canada and the USA are the main importers of frozen durian (Table 15.5), while Hong Kong, Taiwan and Singapore are the main importers of durian paste (Table 15.6). (It is interesting to note that Chinese people prefer durian paste, while Westerners prefer frozen durian.) The promotion of durian products will help familiarise overseas consumers with the fruit and increase the subsequent sales of these products.

Table 15.4: Quantity of durian products exported from Thailand annually (Department of Customs Service, Ministry of Finance, Thailand)

	Frozen durian		Durian paste	
Year	tonnes	US$ million	tonnes	US$ million
1988	581	2.68	55	0.32
1989	1 266	3.82	7	0.03
1990	1 261	4.37	2	0.02
1991	1 572	5.34	1	0.0004
1992	2 390	7.07	7	0.03
1993	2 559	6.99	17	0.04
1994	3 329	8.63	18	0.05
1995	3 547	8.03	6	0.08
1996	5 635	11.26	6	0.03

Table 15.5: Importing countries, quantity and value of frozen durian exported from Thailand (Department of Customs Service, Ministry of Finance, Thailand)

	1995		1996	
Country	tonnes	US$ million	tonnes	US$ million
Australia	534	1.32	3 354	2.12
Canada	672	0.98	491	0.91
France	41	0.22	30	0.19
Germany	4	0.03	6	0.002
Hong Kong	47	0.03	362	0.30
Taiwan	87	0.09	338	0.26
The Netherlands	21	0.11	45	0.19
USA	2 089	5.08	3 354	6.98
Others	52	0.17	125	0.30

Table 15.6: Importing countries, quantity and value of durian paste exported from Thailand (Department of Custom Service, Ministry of Finance, Thailand)

	1995		1996	
Country	kg	US$	kg	US$
Australia	244	1 403.16	1 540	9 011.00
Hong Kong	2 374	11 841.64	2 665	14 809.64
Singapore	3 106	9 053.68	750	4 418.40
Others	318	1 500.24	626	2 527.00

Export market problems

Some of the problems facing Thai export marketers of durian are listed below:

High production costs

These are due to a rapid increase in the price of pesticides, fertilisers and labour. This in turn increases the sale price of fresh fruit and durian products to all parties involved.

Packaging

At present, growers still prefer to export fresh durian which increases the cost of packaging materials and transportation. However, if proper fruit packaging is neglected because of the cost, this will result in post-harvest losses at the export markets. In addition, work must be done on durian product packaging to attract more consumers.

Aroma

Durian consumption is limited mainly to Asian countries. This may be due to the fruit's strong smell and high calorie content. As a result, greater promotion of durian in Europe, America and Africa is necessary.

Storage

Durian has a short shelf-life and cannot be stored for long periods at temperatures below 10°C, without risking chill damage. This makes it very difficult to expand into overseas markets.

Consumer preferences

Consumers of Thai durian are located in Asia, the Middle-East, Europe and America and each prefer different cultivars and quality (Table 15.7). Exporters therefore, must cater to the need of these consumers individually.

Table 15.7: Durian cultivars and quality preferences of importing countries (Department of Trade Promotion, Ministry of Commerce, Thailand)

	Preferences	
Country	*Cultivar*	*Quality*
Developed countries		
Australia, Austria, Canada, Denmark, England, Germany, Japan, Sweden, Switzerland, USA	*Chanee, Mon Thong*	*Half-ripe fruit with weaker aroma and firm*
Asian countries		
Hong Kong, Malaysia, Singapore, Taiwan	*Chanee, Kan Yao*	*Hong Kong and Taiwan prefer fully ripe fruit, while Malaysia and Singapore prefer over-ripe fruit with strong aroma and soft aril*
Middle Eastern countries		
Bahrain, Brunei Darussalam, Saudi Arabia, United Arab Emirates	*Chanee, Kan Yao*	*Similar to developed countries*

Malaysia

Malaysia has a planted area for general fruit production of approximately 123 594 hectares. The main crops grown in this country are banana, durian, jackfruit, langsat, mango, mangosteen, papaya, pineapple, watermelon, rambutan and starfruit, with banana, durian, pineapple and rambutan accounting for 75% of the total fruit production.

Malaysia has 30 360 hectares for durian production and about 50% of this total durian production comes from Perak in northern Malaysia and Johore in the south. The first harvesting season occurs from June to August and the second crop from November to late January. Malaysian fruit appears on the market about two months later than that from Thailand.

Local market

The durian marketing pattern in Malaysia is not as complicated as in Thailand (Figure 15.2). The importer plays an important role in collecting and distributing fruit for the local market via the wholesaler and retailer.

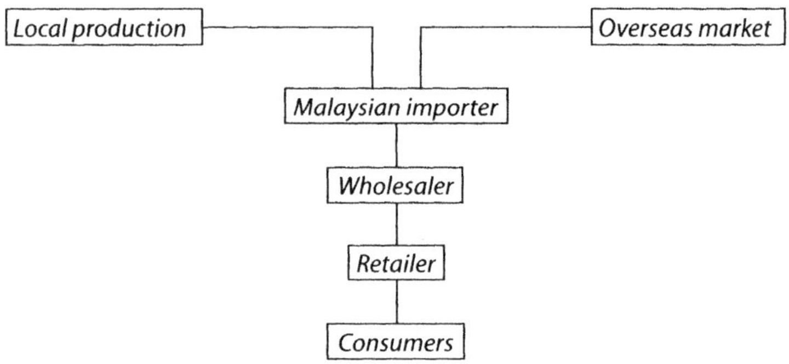

Figure 15.2 *The system for marketing durian in Malaysia*

Although Malaysia produces durian for both local and export markets, it still needs to import additional fruit, mainly from Thailand (Table 15.2) but also from Indonesia and the Philippines. (However, this quantity is still less than it exports.) This may be due to the earlier durian harvesting season in Thailand.

Local market problems

When Thai exporters ship fruit to Malaysia, problems can occur because of the following factors:

Inconsistent sales

Malaysia's own production of durian causes fluctuations in the quantities of fruit imported.

Sale methods

Durian from Thailand is sold on a consignment basis. This limits Thai exporters' bargaining power in price negotiations.

Consumer preferences

Malaysians prefer to eat over-ripe durian. As a result, Thai exporters must adapt the quality of their fruit to account for this customer preference.

Export market

Malaysia exports large quantities of durian to Singapore (Table 15.8). This may be due to their close borders and Singapore's similar preference for over-ripe fruit. Malaysia also exports durian on a smaller scale to Brunei Darussalam, Hong Kong and USA.

Table 15.8: Quantities of Thai and Malaysian durian imported into Singapore ([1] Singapore Trade Statistics; [2] Department of Customs Service, Ministry of Finance, Thailand)

	Malaysia[1]		Thailand[2]	
Year	tonnes	percentage	tonnes	percentage
1979	4 947	84.00	942	16.00
1980	9 740	98.35	163	1.65
1981	9 262	96.76	310	3.24
1982	13 105	97.87	285	2.13
1983	7 841	77.74	2 245	22.26
1984	29 325	98.20	537	1.80
1985	21 567	95.67	977	4.33
1986	26 367	98.52	396	1.48
1987	21 114	93.95	1 360	6.05
1988	32 417	93.98	2 076	6.02
1989	26 058	92.98	1 966	7.02
1990	36 552	99.75	90	0.25

Singapore

Singapore has a small population but one with a high income and a strong purchasing power (ranked second only to Japan among the Asian countries). As mentioned previously, Malaysia is Singapore's primary source of imported durian, with a market share of over 99% (Table 15.8) There are several reasons for this. Firstly, Malaysia's proximity to Singapore reduces the cost of shipment, resulting in lower prices than those from Thailand. Secondly, Singaporean consumers are more familiar with durian from Malaysia which tend to be harvested when over-ripe and have a strong smell, soft aril and large seeds. In addition to Malaysia, Singapore also imports very small quantities of durian from Thailand, the Philippines and Indonesia.

Singapore imports durian from Malaysia twice a year, firstly in small quantities from December to February and secondly, in much larger quantities from May to August. Since the fruit and vegetable market is very competitive in Singapore, importers prefer consignment sales to avoid the risk of price fluctuation. Importers distribute to the wholesale market who then sell the fruit on to retail shops, supermarkets, restaurants and hotels (Figure 15.3). Importers may also sell fruit to processing plants who make durian products.

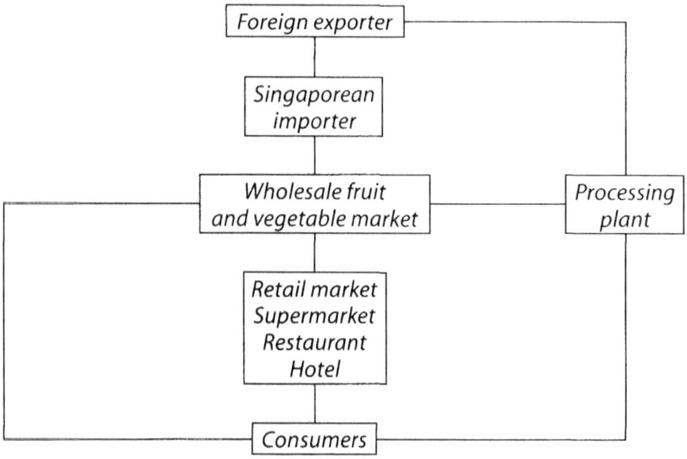

Figure 15.3 *The system for marketing durian in Singapore*

Hong Kong

Hong Kong is one of the main markets for many of Thailand's tropical fruits, the most popular of which include durian, longan, mangosteen and papaya. Approximately 99% of the durian imported into the Hong Kong fruit market comes from Thailand (Table 15.9). Consumers in Hong Kong have a strong purchasing power and are willing to pay for high quality fruit.

Hong Kong imports fruits from all over the world and the durian export period coincides with the importing period of many temperate fruits, such as apples, pears, cherries and strawberries. As a result, these temperate fruits compete with durian in the fruit market. Thailand's maintenance of high quality fruit and reasonable prices must therefore be key factors in attracting consumers in Hong Kong.

As the people of Hong Kong are familiar with durian from Thailand, promotion is less important than retaining the high quality of the fruit. Importers distribute fruit to wholesalers who then sell it on to hotels, restaurants and institutional consumers via retail middlemen (Figure 15.4). Importers may also distribute durian to brokers who sell the fruit to retailers and vendors.

Table 15.9: Quantities of Thai and Malaysian durian imported into Hong Kong ([1] Department of Statistics, Malaysia; [2] Department of Customs Service, Ministry of Finance, Thailand)

	Malaysia[1]		Thailand[2]	
Year	tonnes	percentage	tonnes	percentage
1978	–	–	2 066.7	100.00
1979	–	–	2 818.9	100.00
1980	–	–	3 135.6	100.00
1981	–	–	6 651.0	100.00
1982	1.52	0.08	1 895.8	99.92
1983	3.56	0.18	1 960.8	99.82
1984	2.88	0.14	2 108.0	99.86
1985	6.32	0.08	8 312.6	99.92
1986	4.59	0.09	5 349.1	99.91
1987	5.77	0.07	8 762.4	99.93
1988	2.46	0.06	4 315.3	99.94
1989	1.24	0.02	5 713.7	99.98
1990	9.90	0.09	10 971.1	99.91
1991	5.72	0.10	5 858.5	99.90

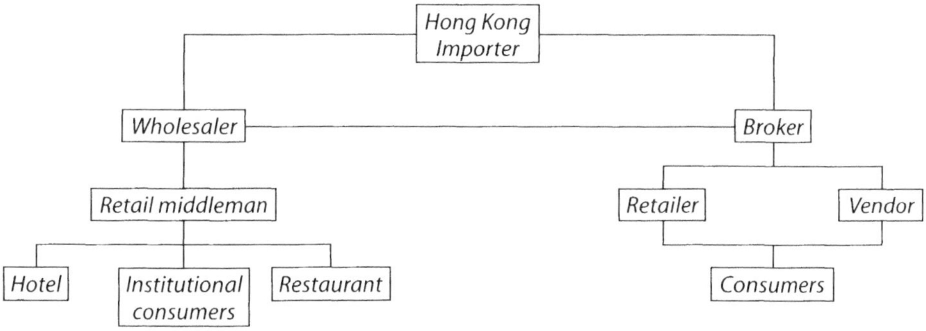

Figure 15.4 *The system for marketing durian in Hong Kong*

Japan

Japan still largely depends on the importation of fruits and vegetables from abroad, with fruits making up only 15% of the total imported produce. Although the majority of fruits imported are from temperate climates, the trend for importing exotic tropical fruits increases every year (although fruits from Thailand make up only 1.8% of the total imported fruits sold in Japan). These include durian, grapes, mangos, rambutan and young coconuts. Although the quantity of Thai durian imported into Japan is very small (Table 15.2), frozen durian is imported all year round and there is no other competition for either fresh or frozen durian. Japanese consumers have a high purchasing power, and therefore vigorous promotion of durian is desirable to familiarise them with the Thai fruit.

All durian imported into Japan are sold in wholesale markets, mostly passing through an auction process. Wholesalers generally distribute the fruit through a second wholesaler to retailers (Figure 15.5). Some fruit, however, will not be auctioned but will be distributed to independent wholesalers who then on-sell to retailers.

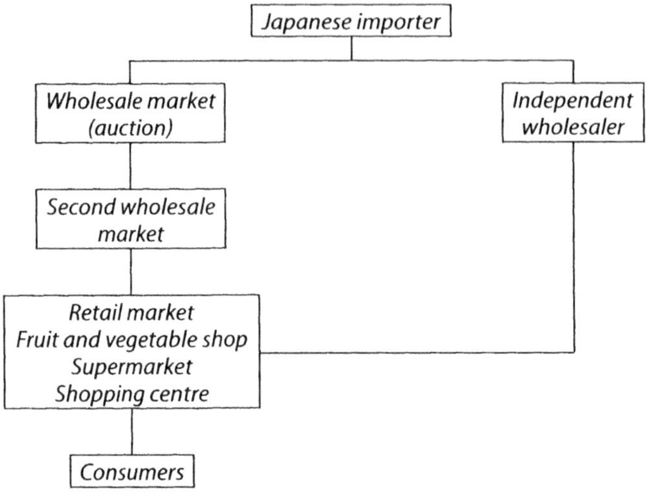

Figure 15.5 *The system for marketing durian in Japan*

Brunei Darussalam

Brunei Darussalam imports a great many tropical fruits including bananas, durian, grapes, guavas, lychees, mangos, mangosteen, pummelo, rambutan, sugar apples and watermelons, with durian being amongst the most popular. The quantity of fruit imported from Thailand increases every year (Table 15.10). However, the value of Thai durian makes up only 4.8% of the total quantity of imported fruits and vegetables sold in Brunei Darussalam. Durian is also imported from Malaysia and Indonesia but these quantities amount only to 0.8%, while Thailand maintains a 78% share of this market.

Importers distribute durian either directly to supermarkets and restaurants (who then sell to consumers), or to a wholesaler who then sells to supermarkets and restaurants. Alternatively, foreign exporters may bypass importers altogether and sell to supermarkets and restaurants directly (Figure 15.6).

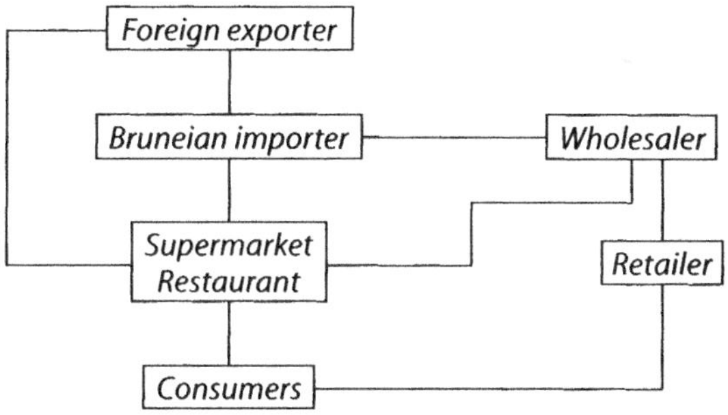

Figure 15.6 *The system for marketing durian in Brunei Darussalam*

Table 15.10: Quantities of Thai durian imported into Brunei Darussalam (Department of Customs Service, Ministry of Finance, Thailand)

Year	Tonnes
1978	0.2
1979	1.2
1980	3.1
1981	62.2
1982	60.2
1983	120.3
1984	100.6
1985	73.0
1986	53.8
1987	112.9
1988	238.0
1989	129.5
1990	109.5
1991	138.6
1992	56.1

Canada and the USA

Both Canada and the USA are becoming increasingly important export markets for Thai durian and durian products, as can be seen in Table 15.11.

Thailand supplies Canada and the USA with the majority of their durian fruit, while smaller amounts are imported from Malaysia. As Canada and the USA both have large populations with high incomes, this means they have a strong purchasing power. However, it must be noted that Canadian and American durian consumers are limited mainly to the Asian immigrant population.

As durian has such a short shelf life and a strong odour, it is difficult to transport by air. In addition, it cannot be stored below 10°C in case of chill damage (Nonthachai, 1994). Sea transport to Canada and the USA is unfeasible because of the vast distances to be covered. These factors mean that frozen durian accounts for a much larger proportion of sales in this market than fresh durian. In fact, the USA imports more frozen durian than any other non-Asian country (Table 15.5).

Table 15.11: Quantities of Thai durian imported into Canada and the USA (Department of Customs Service, Ministry of Finance, Thailand)

Year	Canada (tonnes)	USA (tonnes)
1978	–	1.5
1979	–	3.1
1980	–	2.8
1981	17.9	14.6
1982	17.3	13.9
1983	12.6	35.4
1984	28.4	16.3
1985	50.8	48.5
1986	116.0	273.5
1987	127.0	534.3
1988	32.1	123.7
1989	150.5	129.9
1990	114.5	177.6
1991	158.4	133.6
1992	114.5	15.5

Plate 3.1 Durian flowers growing on branches

Plate 3.2 Cluster of durian flowers

Plate 3.4 Durian fruit on the tree

Plate 4.1 Durian flowers after successful pollination and at the beginning of fruit set

Plate 3.5 Durian fruit showing an edible yellow aril

Plate 5.1 Durian seeds and germinated seedlings

Plate 9.1 Durian flower flushes

Plate 10.1 Stunted seven-year-old tree showing symptoms of macro- and micro-nutrient deficiencies

Plate 9.2 Tying bearing branches to support the fruit load

Plate 10.2 Phytotoxicity of durian foliage associated with 2,4-D

Plate 11.1 Fruit damaged by the fruit-boring caterpillar

Plate 11.2 Early signs of *P. palmivora* foliar blight and seedling dieback

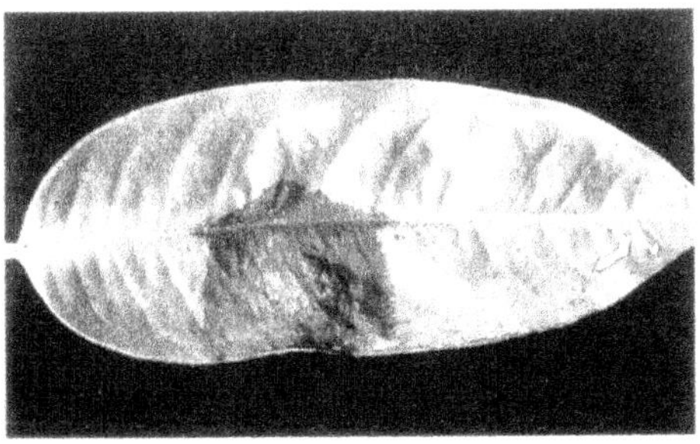

Plate 11.3 Necrotic *P. palmivora* lesions on the leaf

Plate 11.4 *Pythium* root rot extending into the collar tissues

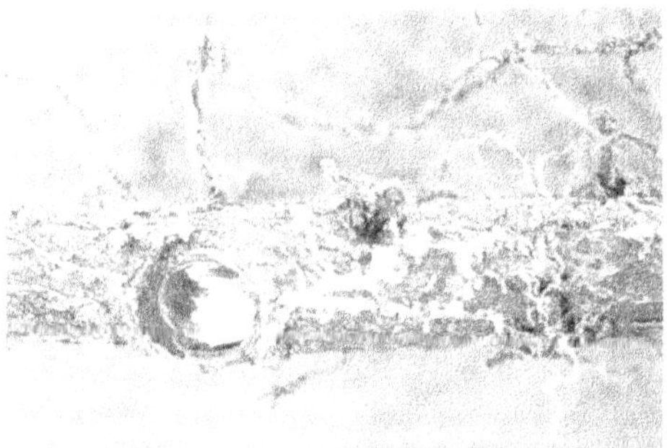

Plate 11.5 Durian root infected with the white rhizomorphs of white root disease

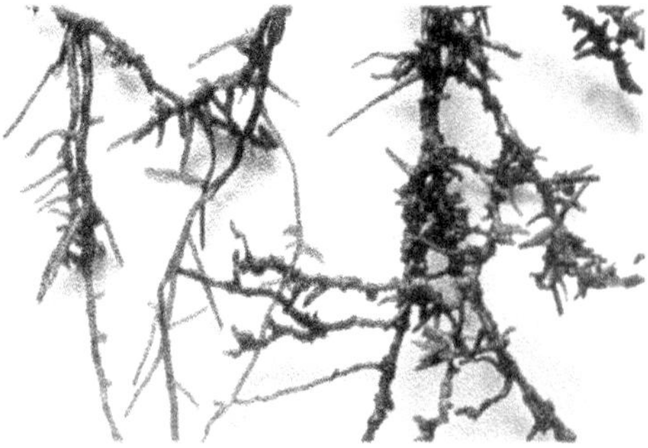

Plate 11.6 *Xiphinema* nematodes: a) infected feeder rootlets on the right and healthy rootlets on the left

Plate 11.6 b) Short stubby feeder rootlets caused by *Xiphinema* nematodes

Plate 11.7 Signs of patch canker disease: a) resinous exudation starting from the crotch region

Plate 11.7 b) Early signs of patch canker trunk disease

Plate 11.8 Early stage of pink disease

Plate 11.9 Velvet felt fungus growing on a twig

Plate 11.12 *Anthracnose* infection in leaves: a) round leaf spot; b) at the leaf apex with characteristic zonate pattern

Plate11.10 *Rhizoctonia* infection in durian leaves: a) early signs

Plate 11.10 b) Advanced signs

Plate 11.11 Leaf spots caused by *Ph. durionis*

Plate 11.13 Fruit rot caused by *P. palmivora*

Plate 11.14 a) Fruit rot caused by *S. rolfsii*

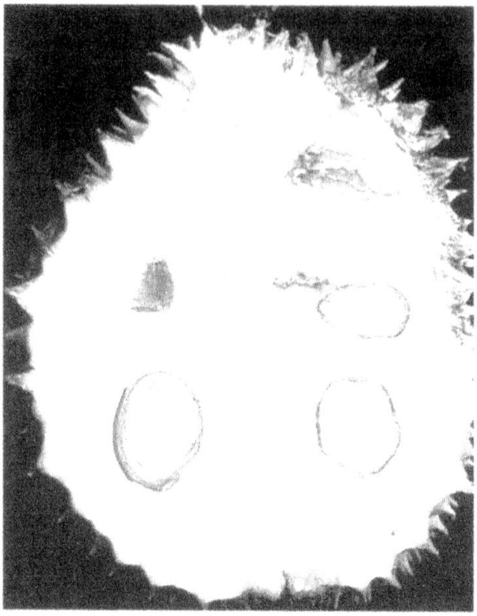

Plate 11.14 b) Note necrosis extending internally into the seeds

Plate 11.15 Necrotic fruit lesion associated with *Rhizopus stolonifer*

Plate 11.16 Rot near the end of the fruit associated with *L. theobromae*

Plate 11.17 Superficial dry rot associated with *Curvularia*

Plate 13.1 a) Harvested durian with stalk intact above abscission point

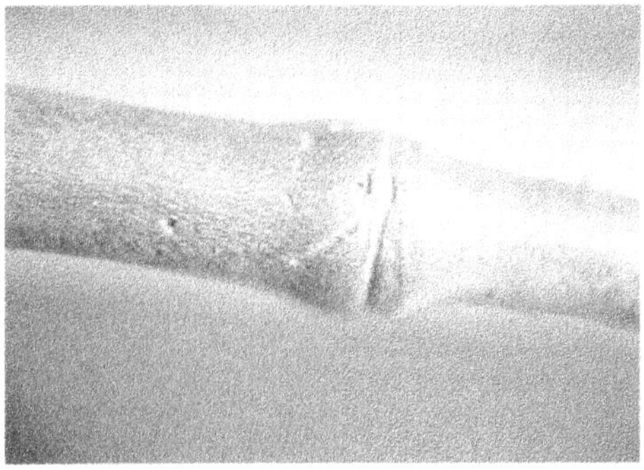

Plate 13.1 b) Abscission point

Plate 13.6 Durian stalks wrapped in banana leaves

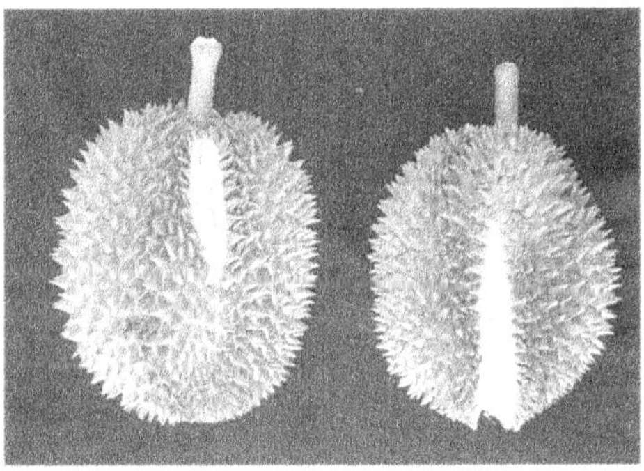

Plate 13.2 Here dehiscence has occurred at the stem end of the fruit (left) and at the blossom end (right)

Plate 13.7 a) Durian laid on the ground on plastic sheets

Plate 13.7 b) Durian in piles at a packing house

Plate 13.8 a) Bulk bins made of wood for use during durian harvesting

Plate 13.8 b) Bulk bins made of metal

Plate 13.9 Bamboo basket containing durian intended for the local market (also used for distant markets within the Asian region)

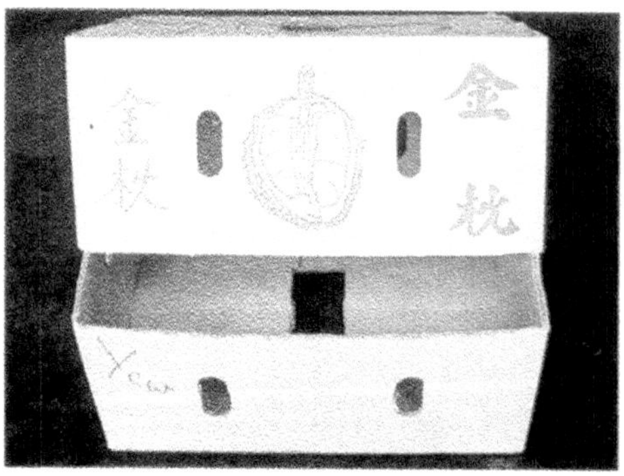

Plate 13.10 Corrugated boxes for durian

Plate 13.11 Durian transported by trucks in bulk

Plate 14.1 Fresh durian aril in polystyrene trays wrapped with PVC film

Plate 14.6 Durian paste wrapped in PVC film

Plate14.8 Durian chips in a can (left) and in a plastic box (right)

Glossary of Terms

A

Abscise:
The dropping of leaf, flower, fruit or other plant parts, usually following the formation of an abscission zone or separation layer, which has a role in the separation (abscisson) of that plant organ from the plant body.

Abscission:
A process in which parts of the plant (leaf, fruit, etc) are detached from the whole plant.

Abscission layer:
A layer of thin-walled cells which are formed at the base of the petiole and facilitate leaf fall.

Acervuli (singular = acervulus):
The sub-cuticular, sub-epidermal, saucer-shaped, cushion-like fruiting bodies of a fungus, bearing condiospores and conidia.

Adjuvant:
A substance added to a chemical spray to enhance the action of the main ingredient.

Air layerage:
A method of stimulating rooting of an attached stem by excluding light, under moist conditions, from a portion of the stem.

Anamorphic state:
Asexual form of a fungus.

Antagonistic soil microflora:
Unicellular plants living in soil that interfere with other kinds of organisms, eg, bacteria, through the creation of unfavourable conditions, eg, by exhaustion of food supply or by production of a specific antibiotic substance.

Anther:
The upper part of a stamen that contains the pollen grains in pollen sacs (anther sacs) ie, the pollen-producing organ of the flower.

Anthesis:
The flowering period or efflorescence. This is the time of full bloom which lasts until fruit set.

Anthracnose:
Disease characterised by black, sunken lesions caused by certain imperfect fungi which produce conidia in acervuli.

Apical dominance:
The characteristic method of branching control which exists in plants where the apex, or top growing point, suppresses the growth of buds close by on the shoot or root.

Apical meristem:
The apex or growing tip of a shoot or root.

Aril:
A fleshy or leathery outer covering of the seed in addition to the seed coat.

ASEAN:
Association of South-East Asian Nations.

Asexual propagation:
Increase in plant numbers by methods other than seedage, ie, clonal or vegetative propagation.

Auxin:
Natural plant growth hormone produced in apical regions. It inhibits growth of lateral buds.

B

ß-galactosidase:
An enzyme which digests galactose.

Basidiomycetes:
Taxonomic group of fungi that produces basidiospores on the basidium.

Basidiospore:
A spore produced on the basidium.

Basidium:
A specialised sexual reproductive cell found in the fertile area of the hymenium of all basidiomycete fungi.

Bipyridilium:
A group of herbicides that inhibit photosynthesis in weeds. The common member in this group is 'paraquat'.

Bract:
A modified leaf, usually associated with a flower or inflorescence.

Broadcasting:
Spreading (sowing) fertilisers over a wide area especially by hand.

Budding:
The asexual propagation process of inserting a bud from a scion shoot into the bark of a rootstock.

Bud patch:
A piece of bud from a scion used in budding.

Bukit:
A Malay word meaning 'hill'.

C

Calcareous soil:
A soil rich in calcium content.

Calyx:
Outermost part of a flower, consisting usually of green, leaf-like membranes known as sepals that in the bud stage enclose and protect the other flower parts.

Carbohydrate:
A chemical containing carbon, hydrogen and oxygen with the latter in a 2:1 ratio. It occurs, for example, in high amounts as starch in seeds and tubers.

Cellulase:
An enzyme which digests cellulose.

Chanee:
A name of one of the leading durian cultivars of Thailand.

Chlamydospores:
An asexual spore found in fungi, primarily for surviving adverse conditions, originating by modification of one or more hyphal segments.

Chloroplast:
A plastid containing chlorophyll and usually other pigments, found in the cytoplasm of plant cells.

Chlorosis:
A condition of plants in which normally green parts become pale green or yellow (chlorotic) due to the prevention of chlorophyll formation resulting from a mineral deficiency or a disease.

Climacteric:
A sudden rise or peak in respiration, cf. climacteric rise, climacteric peak.

Clonal resistance:
A clone of a plant that is resistant to disease or pests.

Clonal selection:
A clone of a plant that has been selected for special characteristics, eg, resistance to disease, pests or adverse conditions. It also refers to the selection of fruit quality, growth habit, etc.

Clone (or clonal):
A series of identical cells or individuals that have developed from a single cell or individual.

Conidia:
Asexual reproductive spores of some ascomycetes and imperfect fungi (singular = conidium).

Corolla:
The collective term for the petals of a flower.

Corymb:
A type of inflorescence.

Corticolous:
An organism living off the surface of living bark, eg, lichen.

Cotyledon:
A seed leaf or first leaf of a fully developed embryo. It is evident below ground in hypogeous seedlings and above ground in epigeous seedlings.

Cross pollination:
Pollination resulting from the transfer of pollen from the anther of one flower to the stigma of a flower from another plant.

Cultivar (cv.):
Term derived from cultivated variety: the cultivar of a plant may be propagated sexually or asexually, in cultivation, to maintain its distinctive characteristics.

D

Dehiscence:
The natural bursting of anthers or other organs to release the contents.

Dipterocarp:
The anglicised name of plant species in the *Dipterocarpus* genus.

Dolomite:
Fertiliser containing calcium and magnesium.

Dolomite limestone:
Soil supplement which is a good source of calcium and magnesium. It is often added to compost piles along with other nutrients to increase the supply of essential nutrients, and to raise the pH.

Dormancy:
The condition of a bud or seed characterised by lack of visible growth.

Durian kuan:
A durian cake in Thai.

Durian pulut:
Common delicacy of durian mixed with glutinous rice. It can also be used to describe the kind of durian that is commonly eaten with glutinous rice.

Dusun:
The home garden system in Malaysia, characterised by growing many kinds of fruit trees in a small area.

E

Ecdyses:
Moulting, usually occurs in insects (singular = ecdysis).

Ectoparasite:
An organism living parasitically on the outside of another organism.

Electrophoresis:
A method of separating substances using the movement of colliodal particles in a fluid under the influence of an electric field.

Embryo:
The rudimentary plant contained in a seed.

Enzyme:
A protein found in living cells which speeds up certain chemical reactions but is not used up in the reaction.

Epicalyx:
A set of bracts close to and resembling a calyx. Sometimes called calycle.

Epidemiology:
The study of the incidence and distribution of diseases, and their control and prevention.

Epigeous:
A seedling whose cotyledons appear above ground following seed germination.

Epiphyte:
A non-parasitic plant (often tropical) which grows upon the stem or leaves of another plant, the former depending on the latter for mechanical support (epi = upon; phyte = plant).

Ester:
An organic compound corresponding to the inorganic salts formed in acid by the replacement of hydrogen by an alkyl radical.

Ethephon:
Common name for 2-chloroethyl-phosphonic acid, which produces ethylene when applied to a plant.

Ethylene:
One of the plant hormones which causes leaf fall (abscission). It is used to induce flowering in pineapple, and to hasten ripening in fruits including durian.

Ex-plant (explant):
Living tissue taken from the natural site of growth, usually placed in a medium or culture.

Exudation:
The process of leaking liquid that in this case is associated with a lethal browning of explants in some woody plant species.

Exude:
To discharge slowly, or leak liquid (exudate) sometimes through pores or cuts but often by diffusion into the medium.

F

Fascicle:
A close tuft of branches (or other plant parts) arising from the same place.

Feeder roots:
Fine roots and root branches with a large absorbing area (root hairs).

Fertilisation:
Fusion of male and female germ cells following pollination.

Fertiliser:
A term referring to a nutrient source such as inorganic fertiliser containing nitrogen (N), phosphorus (P) and potassium (K), or an organic fertiliser such as manure containing high amounts of nitrogen.

Fibrous root:
A much-branched root system with lack of apical dominance.

Filament:
The stalk supporting the anther in a flower.

Flush (of leaf):
A sudden growth of leaf tissue.

Folicolous (of lichen):
Living on leaves.

Fruiting habit:
The location and manner in which fruit is borne on woody plants.

Fruit set:
The persistence and development of an ovary or adjacent tissue following anthesis.

Fungicide:
A substance for destroying fungi or protecting plants from fungi.

Funiculus:
The slender stalk of an ovule.

G

Gene:
Factor of inheritance carried in chromosomes of all cells.

Genera:
Plural of genus.

Genetics:
The science of heredity (gene transmission) and variation.

Genetic erosion:
The loss of the genes in a species, eg, the extinction of a plant species caused by the destruction of forest.

Genus:
Taxonomic term for a group of one or more closely related species, eg, *Durio* is the genus in *Durio zibethinus* (common durian) and *Durio lowianus* (chaarian).

Germinate:
To sprout, of a seed, spore, bud or other structure.

Germination:
The process of growth of a seed, spore, bud or other structure.

Gibberellin:
One of the types of plant hormones produced by vascular plants and some fungi. In vascular plants gibberellins promote cell elongation and often also cell division.

Globose:
Almost spherical.

Grafting:
A method of asexual propagation in which parts of separate individuals are united by matching their tissues so that union and growth can occur.

Grana:
An organelle in chloroplasts – groups of disc-shaped, flattened vesicles (thylakoids) stacked like coins in a pile, vesicle membranes bearing photosynthetic pigments.

Ground cover:
Plants used in lieu of grass for holding soil and giving foliage texture.

Growing season:
For durian, the period between beginning of growth in the rainy season (spring) and the cessation of growth in the beginning of the dry season (winter).

Growth regulator (plant regulator):
A chemical (mainly organic) eg, gibberellin, capable of altering growth characteristics of plants.

H

Hardening or hardening-off:
Subjecting plants to adverse conditions to hasten tissue maturation and increase hardiness.

Heat units:
Number of degrees above a certain base temperature multiplied by a designated number of hours to serve as a guide to maturity.

Herbicide:
A chemical for killing weeds.

Hermaphrodite:
A flower with both a functional stamen and pistil.

Heterothallism:
The type of sexual reproduction where conjugation is possible only through the interaction of different thalli.

Homothallism:
The type of sexual reproduction where conjugation can occur without the interaction of two different thalli.

Hybrid:
Offspring of a cross between two species.

Hybridisation:
The crossing of plants with different genetic constitutions.

Hydrotic:
Water-soaked.

Hyphae:
The threadlike elements of the mycelium of a fungus (singular = hypha).

Hypogeous:
A seedling whose cotyledons remain below ground during seed germination and seedling development.

I

Inarching:
A method of grafting in which one of the partners (stock plant) is decapitated and inserted into the other (scion branch) so that there are two parts below the graft and one above.

Incompatible:
Different kinds or varieties that do not successfully cross-pollinate or intergraft.

Induction (floral):
The condition required for floral initiation.

Inflorescence:
The part of a plant comprising the flowers and flower-bearing stalks.

Initiation (floral):
The first discernible change from a vegetative bud to a floral primordium.

IQF (Individual Quick Freezing):
Process whereby products are individually frozen so that they can be poured freely from a bag without sticking.

Insecticide:
A material for killing insects.

Insect vector:
An insect which transmits a disease.

Integrated pest management (IPM):
An approach to pest control in which various methods are combined to provide long term control.

Intensive production:
A method of increasing the productivity per unit area of land by increasing the expenditure of labour and capital.

Intercropping:
In this context, growing short-term crops between rows of trees in a young orchard for full utilisation of land.

Isozyme or isoenzyme:
Different chemical forms of the same enzyme. Thought to be important in adaptation to environmental extremes.

K

Kampong:
(Malay) A rural village.

L

Lamina:
Leaf blade, expanded portion of the leaf.

Langsat:
A kind of tropical fruit, *Aglaia domestica*, grown commercially in Thailand and the Philippines.

Lateritic (of soil):
A red residual soil in humid tropical and subtropical regions, sometimes known as tropical soil.

Layering:
A method of asexual propagation that involves the formation (naturally or artificially) of adventitious roots from stem sections while they are still attached to the parent plant.

Leaching:
The process of removing soluble minerals from the soil by percolation to the water table.

Leaf axil:
The place where the upper side of a leaf petiole joins the stem.

Lempok:
A durian cake.

Lenticel:
A raised pore in the stem of woody plants which allows gas exchange between the atmosphere and the internal tissues.

Lignicolous (of lichen):
Growing on or in wood.

Loam:
Ideal soil mixture made up of varying parts of sand, silt and clay.

Location:
A climatic or geographic area.

Locule:
A compartment, chamber or cavity, eg, in fruit, ovary or anther.

Longan:
A kind of tropical fruit, *Dimocarpus longan*, commonly cultivated in South-East Asia and China.

M

Mangosteen:
A kind of tropical fruit, *Garcinia mangostana*, also known as the 'Queen' of tropical fruit.

Marcotting:
Synonym of 'air layering' ie, rooting a girdled stem while still attached to the plant.

Maturity:
The state of ripeness. Usually the stage of development which results in maximum quality.

Meristem:
A tissue with cells which are capable of active cell division and differentiation into specialised tissues.

Monoculture:
Growing of a single crop in the area.

Monophyletic:
Individuals descended from a common ancestor.

Mulch:
Material, usually organic, spread around or over a plant to retain moisture, enrich or insulate the soil.

Mycelium:
A mass of hyphae (thallus) making up the fungus body.

N

Necrosis:
Death or decay of tissue.

Nematicide:
A chemical that inhibits or kills nematodes.

Nematode:
A microscopic, non-segmented round worm found in soil and other habitats. Many species attack plants and animals.

N-P-K:
Refers to a nitrogen-phosphorus-potassium ratio used for inorganic fertilisers, such as a 10-10-10 fertiliser (10 parts N, 10 parts P and 10 parts K).

Nursery:
An establishment for growing, handling, or retailing woody and herbaceous plants.

Nutrients:
Elements available through soil, air, and water, which the plant utilizes in growth.

O

Oil palm:
A kind of palm tree, *Elaeis quineensis*, economically grown in Thailand and Malaysia for 'palm oil'.

Open pan evaporation:
An instrument for measuring evaporation rate of water in the field.

Organic matter:
Plant and animal residues and remains.

Ovary:
The enlarged basal portion of the pistil, in which the seeds are borne.

Oviposition:
Laying of eggs by insects.

Ovule:
The part of the ovary of a plant containing the egg cell or female gamete which, after fertilisation, develops into a seed.

P

Panicle:
A type of inflorescence (flower) in which the main axis has several alternate or spirally-formed branches (racemes) each bearing one or more flowers.

Pan:
A shallow, wide, open container, usually of metal and without a lid, used for holding rain water or other liquid, to study the evaporation rate.

Paraquat:
A herbicide in the bipyridilium group.

Pathogen:
A disease-causing organism.

Pectinesterase:
An enzyme which catalyses the digestion or breakdown of pectin.

Pedicel:
The stalk bearing a single flower of an inflorescence.

Peduncle:
The stalk supporting either an entire inflorescence or a solitary flower.

Peltate:
In the form of a round plate with a stalk in the centre on the underside.

Perfect flower:
Flower possesing both functioning stamens and pistils ie, a hermaphrodite flower.

Petiole:
The stalk of a leaf.

Phloem:
Food conducting tissue in a vascular plant located just inside the bark of woody plants.

pH scale:
A means of expressing acidity or alkalinity with a range of 1 to 14. A reading of 7 is neutral (ie, distilled water), below 7 is acid, and above 7 is alkaline.

Photosynthesis:
The process by which the energy of sunlight is used by green plants to synthesize carbohydrates from carbon dioxide and water.

Pistil:
The female part of the flower comprising the stigma, style and ovary.

Pollination:
Transfer of pollen from an anther to a stigma.

Polygalacturonase (PG):
An enzyme that digests part of the pectin that forms plant cell walls. PG causes ripening fruit to become soft.

Pre-climacteric:
A period of minimal respiration before a sudden rise (or peak) in respiration.

Processing:
Preparing for future use as in canning, freezing, preserving and dehydrating.

Primordium:
An organ, structure or tissue in the earliest stage of development.

Productive stage:
The period of fruit-bearing productivity.

Pruning:
The selective removal of leaves, shoots, twigs, branches, or roots of a plant to increase its productivity.

Pubescent:
Hairy or downy coverage.

Pulan-buah:
An island full of fruits.

Pycnidium:
Asexual, flask-shaped or globose fungal fruiting structure bearing conidia.

Pycnospores:
Spores or conidia formed in pycnidia.

Pyriform:
Pear-shaped.

Q

Quality:
Various factors including flavour, texture, appearance, odour, and food value which influence judgement of produce.

Quarantine:
Regulation forbidding sale or shipment of plants, usually to prevent disease or insect invasion of an area.

Quincunx:
An arrangement of five objects such as trees, with one at each corner of a rectangle and one at the centre.

R

Rambutan:
A tropical fruit tree, *Nephelium lappaceum*, of South-East Asia, bearing edible oval red fruit with soft spines.

Receptacle:
The enlarged end of the pedicel or peduncle to which other flower parts are attached.

Resistance:
In this context, the ability to suppress or retard the development of a pathogen or other injurious factor.

Respiration:
Oxidation of foods by plants and animals for energy release. Carbon dioxide and water are liberated.

Rhizomorph:
A densely packed strand of fungal hyphae resembling a root.

Rhizosphere:
The soil around a living root.

Ripeness:
State of complete maturation before breakdown (see also Maturity).

Riverine:
The banks of a river.

Rootstock:
Plant material with good rooting properties on which a scion with desirable vegetative and fruiting characteristics is grafted or budded.

S

Scion:
A detached piece of stem (bud or budstick) inserted into another plant in budding or grafting.

Self-fertilisation:
The fertilisation of the ovule of a variety or flower type by its own pollen.

Self-pollination:
Pollination resulting from the transfer of pollen from the anther of a flower to the stigma of the same flower or tree.

Sepal:
One of the flower parts, usually green and leaf-like.

Sexual propagation:
The means of increasing plant numbers by sexual reproduction and the production of seeds.

Sink:
The actively growing organs in a plant (eg, apex, root, young fruit, etc) that have a strong demand for food (eg, carbohydrates) for their growth.

Soluble powder:
Pesticide formulation in the form of a powder to which a wetting agent has been added.

Source:
The organ in plants (eg, a mature leaf) that manufactures food (eg, carbohydrates) and sends it to the 'sink'.

Species:
A taxonomic term to describe a group of closely related individuals that are potentially capable of interbreeding. A subdivision of a genus.

Sporangia:
A group of sac-like fungal fruiting structures whose contents are differentiated into spores (singular = sporangium).

Stomata:
Small opening generally in the lower epidermis of leaves and in some stems through which gases pass (singular = stoma).

Stroma:
a) A mass of vegetative fungal tissue or a mixture of host (substrate) and fungal tissue in or upon which the reproductive body develops. b) A colourless matrix of chloroplasts in which the grana are embedded.

Subgenus:
Taxonomic term to describe the subdivision of a genus, each subgenus contains one or more species that are more similar to each other than to any other species in the genus.

Surfactant:
An active surface agent, such as a detergent, that aids the penetration of foliar applied chemicals.

Systemic fungicide (or insecticide):
A fungicide (or insecticide) which is absorbed by the plant and moves in the sap to all parts of the plant.

T

Tempoyak:
A fermented durian paste.

Terminal leafbud:
The leaf bud at the apex of the stem.

Thalli:
Plant bodies such as fungi, mosses etc, without true roots, stems or leaves (singular = thallus).

Thorax (in insects):
The group of three segments behind the head which bears the three pairs of legs and (when present) the wings.

Tolerance:
Ability to survive and produce well in the presence of a destructive pathogen, pest or adverse environmental conditions.

Top-working:
Changing the variety of a tree by inserting buds or grafts on its branches.

Translocation:
The movement of water and other dissolved substances through the conducting system of the plant.

Transpiration:
The loss of water vapour from plant surfaces, chiefly through the stomata of leaves.

V

Variety:
A group of closely related plants of common origin which have similar characteristics having been bred for a specific feature or group of features.

Vector:
Any organism that is a carrier of a pathogen.

Velutinous:
Covered with dense, soft, silky hairs ie, having a velvety texture.

W

Wilting:
A condition in plants in which cells lose their turgidity, and leaves and young stems etc, droop as a result of excessive water loss (in transpiration).

Z

Zonate:
Having belted, striped or ringed zones.

Bibliography

Abang Mohd Mokhtar, A.P. 1991, The taxonomy of *Durio* in Sarawak (Malaysia), M.Sc. thesis, University of Reading, Reading, England.

Abeles, F.B., Morgan, P.W. and Saltveit Jr, M.E. 1992, *Ethylene in Plant Biology*, Academic Press Inc, New York, USA.

Alameyda, N., Malo, S.E. and Martin, F.W. 1979. *Cultivation of Neglected Tropical Fruits with Promise. Part 6 – The Rambutan.* USDA/United States Department of Agriculture, Science and Educational Administration, Beltsville, Maryland, USA, pp. 1–4.

Atta-Aly, M.A. 1992, Effects of high temperature on ethylene biosynthesis in tomato fruit, *Post-harvest Biological Technology* 21: 19–24.

Baldry, J., Dougan, J. and Howard, G.E. 1972, Volatile flavouring constituents of durian, *Pytochemistry* 11: 2081–4.

Barrett, O.W. 1912, The durian, *Philippines Agricultural Review* 5: 589–92.

Bertanam pohon buah-buahan 2, 2nd ed., Penerbitan Yayasan Kanisius, Yogyakata, Indonesia (in Indonesian).

Bhusiri, S. 1987, *The Story of Durian*, Ratana Farm, Trang, Kao Chong, Thailand (in Thai).

Biale, J.B. 1960, The after-harvest biochemistry of tropical and subtropical fruits, *Advances in Food Research* 10: 293–354.

Boonbongan, C. 1956. Improvement of native durian varieties. *Kasikorn* 29(3): 241–245 (in Thai).

Booncherm, P. 1990, Physiological changes in durian (*Durio zibethinus* Murr.) (cv.) Chanee after harvest, M.Sc. thesis, Kasetsart University, Bangkok, Thailand (in Thai with English abstract).

Boonyakom, C. 1995, Knowledge of durian, Kasetsart University, Bangkok, Thailand (in Thai).

Brown, W.H. 1954, Useful plants of the Philippines, Technical Bulletin No.10, Volume 2, Republic of the Philippines Department of Agricultural National Research, Bureau of Printing, Manila, Philippines.

Buescher, R.W. 1979, Influence of high temperature on physiological and compositional characteristics of tomato fruits, *Leben Wisser Technologie* 12: 162–4.

Bureau of Agricultural Economics 1977, Crop statistics, Ministry of Agriculture, Manila, Philippines.

Burton, W.G. 1982, *Post-harvest Physiology of Food Crops*, Longman, London.

Chaiprasart, P. 1982, Effects of low temperature and fruit coatings ('FMC Sta-fresh 7055' and 'Semperfresh') on the storability of durian, Graduate Special Problem, Department of Horticulture, Kasetsart University, Bangkok, Thailand (in Thai with English abstract).

Chaiprasart, P. 1993, Enzymes and intermediates in ethylene production of durian fruit, M.Sc. thesis, Kasetsart University, Bangkok, Thailand (in Thai with English abstract).

Chaiwongkiat, D. 1992, *Durian's 35*, Aksorn Publishers, Bangkok, Thailand (in Thai).

Chan, H.T. 1986a, Effects of heat treatments on the ethylene-forming enzyme system in papayas, *Journal of Food Science* 51: 551–3.

Chan, H.T. 1986b, Heat inactivation of the ethylene-forming enzyme system in cucumbers, *Journal of Food Science* 51: 1491–3.

Chan, Y.K., Hassan, S., Jamaluddin, S.H. *et al.* 1981, A report on the techno-economic survey of the Malaysian fruit industry 1980, M. Yusof Hashim and M. Ariff Hussein (eds), Malaysian Agricultural Research and Development Institute (MARDI), University of Pertanian, Kuala Lumpur, Malaysia.

Chantaburi Horticultural Research Centre 1996, Delicious dishes from raw durian aril, Horticultural Research Institute, Bangkok, Thailand (in Thai).

Chartjareon, L. 1961, Characteristic studies on four durian cultivars, B.Sc. thesis, Faculty of Agriculture, Kasetsart University, Bangkok, Thailand (in Thai).

Cockburn, P.F. 1976, *Trees of Sabah*, Volume 1, Forest Department, Kum Co. Ltd, Sabah, Malaysia, pp. 20–32.

Coronel, R.E. 1966, Methods of emasculation and pollination of durian flowers, *Agriculture Los Banos* 6 (1): 5–8.

Coronel, R.E. 1977, The growing of durian, Horticultural Report, Extension Circulation No. 16, College of Agriculture, University of the Philippines, Los Banos, Laguna, Philippines.

Coronel, R.E. 1983, *Promising Fruits of the Philippines*, College of Agriculture, University of the Philippines, Los Banos, Laguna, Philippines.

Coronel, R.E. and Manlisis, S.S. 1972, Iron chlorosis in seedlings of fruit trees, College of Agriculture, University of the Philippines, Los Banos, Laguna, Philippines.

Daengkanit, T. 1995a, Effect of temperature on ripening and eating quality of durian (cv.) Mon Thong, Graduate Speciàl Problem, Department of Horticulture, Kasetsart University, Bangkok, Thailand (in Thai).

Daengkanit, T. 1995b, Relationship between activities of polygalacturonase, pectin methylesterase and cellulases and softening of durian pulp during ripening, M.Sc. thesis, Kasetsart University, Bangkok, Thailand (in Thai with English abstract).

Department of Agricultural Extension 1990, Fruit and tree crop production data in Thailand, Fruit Crop Division, Bangkok, Thailand (in Thai).

Department of Agricultural Extension 1992, Fruit and tree crop production data in Thailand, Fruit Crop Division, Bangkok, Thailand (in Thai).

Director of Agriculture 1988, Brunei Darussalam agricultural research in perspective, Brunei Darussalam Department of Agriculture, Ministry of Development, Negara, Brunei Darussalam.

Director General of Food, Crops and Horticulture 1994, Area planted, yield and production of horticultural crops in Indonesia in 1992, Jakarta, Indonesia (in Indonesian).

Durian: it tastes like heaven, it stinks like hell, *Time Magazine* 237 (7): 13.

Eaks, I.L. 1978, Ripening respiration and ethylene production of 'Hass' avocado fruits at 20°C and 40°C, *Journal of the American Society for Horticultural Science* 103: 576–8.

Esau, K. 1977, *Anatomy of Seed Plants*, John Wiley & Sons, New York, USA.

Farmers Association of Malaysian Agriculture 1988/89, Kajian penggunaan hasil-hasil pertanian, Survey Report No. 19, Farmers Association of Malaysian Agriculture, Kuala Lumpur, Malaysia (in Malay).

Galán Sauco, V. 1989, Litchi cultivation, Food and Agriculture Organization Plant Production and Protection Paper 83, Food and Agriculture Organization, Rome, Italy.

Galang, F.G. 1955, *Fruit and Nut Growing in the Philippines*, AIA Printing Press, Malabon Rizal, Manila, Philippines.

Haruthaitanasan, V. 1987, Production of durian powder for food industry, *Kasetsart News* 31 (6): 3 (in Thai).

Hasan, M. and Karim, M.Z. 1990, Application of paclobutrazol to lengthen fruiting season in durian (*Durio zibethinus*). In *Off-season Production of Horticultural Crops*, Book Series No. 41, Food and Fertilizer Technology Centre for the Asian and Pacific Region, Taipei, Taiwan, pp. 89–92.

Hiranpradit, H., Chandraparnik, S. and Salarpetch, S. 1998. *Technology for producing durian.* Kasetsart University Press, Bangkok (in Thai).

Hiranpradit, H., Jantrajoo, S., Lee-ungkulasatian, N. *et al.* 1992, Group characterization of Thai durian, *Durio zibethinus* Murr., *Acta Horticulturae* 321: 263–9.

Hiranpradit, H., Lee-ungulasatian, N., Chandraparnik, S. *et al.* 1992, Quality standardization of Thai durian, *Durio zibethinus* Murr., *Acta Horticulturae* 321: 695–704.

Huan, T.L. 1971, Studies on *Phytophthora palmivora*, the causal organism of patch canker disease of durian, *Malaysian Agricultural Journal* 48 (1): 1–9.

Inthong, P. 1964, A survey of durian varieties at Amphur Muang Nonthaburi, B.Sc. thesis, Faculty of Agriculture, Kasetsart University, Bangkok, Thailand (in Thai).

Jarimopas, B., Siriatchatapong, P., Sukharom, S., *et al.* 1992, Durian sizing machine, *Kasetsart Journal* (Natural Science Supplement) 26: 65–74.

Kaiviparkbunyay, K. 1989, Effect of paclobutrazol on flowering of durian (cv.) Chanee, M.Sc. thesis, Kasetsart University, Bangkok, Thailand (in Thai with English abstract).

Kanapathy, K. 1976, Guide to fertilizer use in Peninsular Malaysia, Bulletin 143, Ministry of Agriculture and Rural Development, Kuala Lumpur, Malaysia.

Kanzaki, S., Yonemori, K., Sugiura, A. *et al.* 1998, Phylogenetic relationships of the common durian (*Durio zibethinus* Murr.) to other edible fruited *Durio* species by RFLP analysis of an amplified region of cpDNA, *Journal of Horticultural Science and Biotechnology* 73 (3): 317–21.

Kasiolarn, H. 1991, Post-harvest physiology of sugar apple (*Annona squamosa* Linn.) and storage of antemoya (*Annona x* 'African Pride') fruit, M.Sc. thesis, Kasetsart University, Bangkok, Thailand (in Thai with English abstract).

Ketsa, S. 1990, Effect of fruit size on weight loss and shelf life of tangerines, *Journal of Horticultural Science* 65: 485–8.

Ketsa, S. 1993, Ripening. In J. Siripanich, S. Ketsa, S. Sangchote *et al.* (eds), Technical development and economic assessment and disease control, storage and ripening of durians for export, Research, Development and Economic Project, Final Report, Kasetsart University, Bangkok, Thailand, pp. 42–77.

Ketsa, S. and Daengkanit, T. 1998a, Changes in softening enzymes of durian fruit ripening, *Acta Horticulturae* 464: 451–4.

Ketsa, S. and Daengkanit, T. 1998b, Physiological changes during post-harvest ripening of durian fruit (*Durio zibethinus* Murr.), *Journal of Horticultural Science and Biotechnology* 73: 575–9.

Ketsa, S. and Daengkanit, T. 1999. Firmness and activities of polygalacturonase, pectinesterase, B-galactosidase and cellulase in ripening durian harvested at different stages of maturity. *Scientia Hortic.* 80: 181–188.

Ketsa, S. and Pangkool, S. 1994, The effect of humidity on ripening of durian, *Post-harvest Biological Technology* 4: 159–65.

Ketsa, S. and Pangkool, S. 1995a, Effect of temperature and humidity on the ripening of durian fruit, *Journal of Horticultural Science* 70: 827–31.

Ketsa, S. and Pangkool, S. 1995b, Ripening behaviour of durian at different temperatures, *Tropical Agriculture* 72: 141–5.

Ketsa, S. and Pangkool, S. 1996, Effect of maturity stages and ethylene treatment on ripening of durian fruit. In C. Frisina, K. Mason and J. Faragher (eds), Proceedings of Australasian Post-harvest Horticulture Conference, Science and Technology for the Fresh Food Revolution, Melbourne, Australia, pp. 67–72.

Khunprom, C. 1995, Post-harvest Research Unit, Central Laboratory and Greenhouse Complex Centre, Kasetsart University, Kamphaengsaen Campus, Nakhon Pathom, Thailand (personal communication).

Kietkong, P. 1957, A survey of durian varieties in Nonthaburi, B.Sc. thesis, Faculty of Agriculture, Kasetsart University, Bangkok, Thailand (in Thai).

Kosittrakun, M. and Nanthachai, S. 1994, Packaging and transportation. In S. Nanthachai (ed.), *Durian: Fruit Development, Post-harvest Physiology, Handling and Marketing in ASEAN*, ASEAN Food Handling Bureau, Kuala Lumpur, Malaysia, pp. 97–105.

Kosiyachinda, S., Nanthachai, S., Kosittrakun, M. *et al.* 1995, Durian fruit: Harvest and post-harvest management, Department of Agriculture, Bangkok, Thailand (in Thai).

Kostermans, A.J.G.H. 1958, The genus *Durio* Adans (Bombacaceae), *Reinward* 4: 357–460.

Krishnamurthi, S. and Rao, V.N.N. 1965. Fruits of the world in Madhras – introductions over a century and their performances. In S Krishnamurthi (ed.), Advances in agricultural sciences and their applications. *The Madras Agricultural Journal*, Corimbatore, India, pp. 421–443.

Lim, T.K. 1993. *Durian diseases and disorders.* 2nd printing. Tropical Press Sdn Bhd, Kuala Lumpur.

Lim, T.K. 1990, *Durian: Diseases and Disorders*, Tropical Press Co. Ltd., Kuala Lumpur, Malaysia.

Lim, T.K. and Chan, L.G. 1986, Parasitism of *Phytophthora palmivora* by *Gliocladium roseum*, *Zeitschrift für Pflanzenkranheiten und Pflanzenschuft* (*Journal of Plant Disease and Protection*) 93 (5): 509–14.

Lim, T.K., Ng, C.C. and Chin, C.L. 1987, Etiology and control of durian foliar blight caused by *Rhizoctonia solani*, *Annals of Applied Biology* 110 (2): 301–7.

Loh, C.L. and Lee, C.S. 1966. Durian variety trial. Malaysian Ministry of Agriculture and Cooperatives Departmental Conference Information Paper 52.

Luardo, B.R. 1972. Durian (Durio zibethinus Murr., Bombacaceae). In *Cultural directions for Philippines agricultural crops.* Vol 1 (Fruit): 113–117. Publ. Aff. Off. Press, Bur. Plant Indus., Manila.

Lurie, S. and Klein, J.D. 1990, Heat treatment of ripening apples: Differential effects on physiology and biochemistry, *Physiologia Plantarum* 78: 181–6.

Malo, S.F. and Martin, F.W. 1979, *Cultivation of Neglected Tropical Fruits with Promise: Part 7 – The Durian*, United States Department of Agriculture, Science and Educational Administration, Beltsville, Maryland, USA.

Martin, F.W. 1980, Durian and mangosteen. In S. Nagy, and P.E. Shaw (eds), *Tropical and Subtropical Fruits: Composition, Properties and Uses*, AVI Publishing Inc, Westport, Connecticut, USA, pp. 407–14.

Mathur, P.B. and Srivastava, H.C. 1954, Preliminary experiments on the cold storage of durian (*Durio zibethinus* Murr.), Bulletin No. 3, The Central Food Technology Research Institute, Mysore, India.

Ministry of Agriculture, Acreages of miscellaneous crops, 1993, Kuala Lumpur, Malaysia.

Mohamed, S. 1990, Extending the shelf life of fresh durian (*Durio zibethinus*), *ASEAN Food Journal* 5: 117–19.

Molesworth-Allen, B. 1965, *Malayan Fruits*, Eastern Universities Press, Singapore.

Monsalud, M.R., Tongacan, A.L., Lopez, F.R. *et al.* 1966, Edible wild plants in the Philippines forest, *Philippines Journal of Science* 95 (4): 431–561.

Mortensen, E. and Bullard, E.T. 1964, *Handbook of Tropical and Subtropical Horticulture*, Department of State, Agency for International Development, Washington D.C., USA.

Namuco, L.O. 1978, Self-incompatibility and seed development in durian (*Durio zibethinus* Murr.), M.Sc. thesis, University of the Philippines, Los Banos, Laguna, Philippines.

Nanthachai, S. (ed.), 1994, *Durian: Fruit Development, Post-harvest Physiology, Handling and Marketing in ASEAN*, ASEAN Food Handling Bureau, Kuala Lumpur, Malaysia.

Nanthachai, S., Siripanich, J., Wahab, A.R. *et al.* 1994, Harvesting indices and harvesting. In S. Nanthachai (ed.), *Durian: Fruit Development, Post-harvest Physiology, Handling and Marketing in ASEAN*, ASEAN Food Handling Bureau, Kuala Lumpur, Malaysia, pp. 77–88.

Navaratnam, S.J. 1966, Patch canker of the durian tree, *Malaysian Agricultural Journal* 45 (3): 291–4.

Ng, S.K. and Thamboo, S. 1967, Nutrient removal studies on Malayan fruits: Durian and rambutan, *Malaysian Agricultural Journal* 46 (2): 164–82.

Ochse, J.J., Soule Jr, M.J., Dijkman, M.J. *et al.* 1961, *Tropical and Subtropical Agriculture*, Macmillan, New York, USA.

Padolina, F. 1931. Vegetative propagation experiments and seed germination. *Philippines Journal of Agriculture* 2(4): 347–355.

Pantastico, E.B., Lam, P.F., Ketsa, S. *et al.* 1984, Post-harvest physiology and storage of mango. In D.B. Mendoza Jr, and R.B.H. Wills (eds), *Mango: Fruit Development, Post-harvest Physiology and Marketing in ASEAN*, ASEAN Food Handling Bureau, Kuala Lumpur, Malaysia, pp. 39–52.

Paveekarn, S. 1988, Prolongation of durian cake by frozen storage, Horticultural Research Institute, Department of Agriculture and Cooperatives, Bangkok, Thailand (in Thai).

Polachart, P. 1952, Durian plantation in Thonburi, *Kasikorn* 25 (6): 505–14 (in Thai).

Polprasid, P. 1960, Durian flowers, *Kasikorn* 33 (2): 37–45 (in Thai).

Polprasid, P. 1967, How to improve fruit set in durian, *Kasikorn* 40 (3): 229–37 (in Thai).

Polprasid, P. 1969, Notes on various aspects of durian pollen grain and fruit set problems, *Kasikorn* 42 (2): 143–54 (in Thai).

Polprasid, P. 1983, *Artocarpus* and *Durio*: South-East Asia regional training course on characterization and preliminary evaluation of plant germplasm, 16–29 October 1983, National Corn and Sorghum Research Centre, Pakchong, Nakhon Ratchasima, Thailand.

Pongsrihadulchai, A. 1984, Survey report on fruit production and marketing in Thailand, Symposium on fruit production and marketing, 9–14 July 1984, Pattaya, Cholburi Province, Thailand.

Praditdoung, S. 1986, Cold storage of durian, *Kasetsart Journal* (Natural Science Supplement) 20: 44–9 (in Thai with English abstract).

Praditkul, U. 1987, Rind anatomy of durian (cv.) Kan Yao, Chanee and Mon Thong, Undergraduate Special Problem, Department of Horticulture, Kasetsart University, Bangkok, Thailand (in Thai).

Punsri, P. 1972, Wild durian, *Puech Suan* 8 (2): 17–22 (in Thai).

Ratanachinakorn, B. and Nanthachai, S. 1994, Sorting and grading. In S. Nanthachai (ed.), *Durian: Fruit Development, Post-harvest Physiology, Handling and Marketing in ASEAN*, ASEAN Food Handling Burcau, Kuala Lumpur, Malaysia, pp. 89–96.

Rattanapruk, V. 1969, Durian propagation by top-working and T-budding, *Kasikorn* 42 (1): 49–55 (in Thai).

Razak, A.R. 1988, Ecology and management of plant parasitic nematodes on fruit in Malaysia, Seminar on Tropical Fruit Cultivation, Pest and Disease Management, 20–21 October 1988, Kota Kinabalu, Sabah State, Malaysia.

Reksodihardjo, W.S. 1962, The species of *Durio* with edible fruits, *Economic Botany* 16: 270–82.

Sangchote, S. 1993, Disease control. In J. Siripanich, S. Ketsa, S. Sangchote *et al.* (eds), Technical development and economic assessment on disease control, storage and ripening of durian for export, Research, Development and Economic Project, Final Report, Kasetsart University, Bangkok, Thailand, pp. 8–15.

Sankhariksha, S. 1968. A comparative study on the different budding techniques in propagating durian. B.S.A. thesis, College of Agriculture, University of the Philippines, Los Banos, Laguna, Philippines.

Siew, L.B., Kosittrakun, M. and Vichittrananda, S. 1994, Pathology and disease control. In S. Nanthachai (ed.), *Durian: Fruit Development, Post-harvest Physiology, Handling and Marketing in ASEAN*, ASEAN Food Handling Bureau, Kuala Lumpur, Malaysia, pp. 62–6.

Singh, K.G. 1980, A Check List of Host and Disease in Malaysia, Ministry of Agriculture, Kuala Lumpur, Malaysia.

Siripanich, J. 1993, Storage. In J. Siripanich, S. Ketsa, S. Sangchote *et al.* (eds), Technical development and economic assessment on disease control, storage and ripening of durian for export, Research, Development and Economic Project, Final Report, Kasetsart University, Bangkok, Thailand, pp. 16–41.

Siripanich, J. and Jarimopas, B. 1993, Post-harvest loss of durian, *National Agricultural Machinery Centre Newsletter* 6 (4): 1–3

Siripanich, J., Ketsa, S., Chana, C. *et al.* 1990, Technical development and economic assessment on disease control, storage and ripening of durian for export, Research, Development and Economic Project, Semi-Annual Report No. 1, Kasetsart University, Bangkok, Thailand.

Smitinand, T. 1980, *Thai Plant Names (Botanical Names–Vernacular Names)*, Royal Forest Department, Bangkok, Thailand.

Soeseno, S. 1968, *Buah-Buah Dikebun Rumah*, P.T. Kinta, Jakarta, Indonesia.

Somsri, S. 1987, Studies on hand-pollination of durian (*Durio zibethinus* L.) (cv.) Chanee and Kan Yao by certain pollinizers, M.Sc. thesis, Kasetsart University, Bangkok, Thailand (in Thai with English abstract).

Somsri, S. 1988, Durian cultivars and cultivation. In Proceedings of the Seminar on Durian, 25–26 February 1988, Thailand Institute of Scientific and Technological Research, Bangkok, Thailand, p. 5.

Srikumlaithong, S. and Swatdita, A. 1990, Elimination of durian odour for export, *Agricultural Science Society of Thailand Newsletter* 13: 431–46 (in Thai with English abstract).

Sriyook, S. 1990, Effect of coating materials on ripening and dehiscene in durians (cv.) Chanee, M.Sc. thesis, Kasetsart University, Bangkok, Thailand (in Thai with English abstract).

Sriyook, S., Siriatiwat, S. and Siripanich, J. 1994, Durian fruit dehiscence: Water status and ethylene, *HortScience* 29: 1195–8.

Stanton, W.R. 1966, The chemical composition of some tropical food plants: IV Durian, *Tropical Science* 8 (1): 6–10.

Subhadrabandhu, S. 1993, Current status and future strategies in durian and mangosteen research in Thailand. In Prosiding Symposium Buah-buahan Kebangsaan Ketiga, Ke Arah Industri Buah-buahan Berdaya Maju, 24–26 September 1991, Genting Highlands, Pahang Darul Makmur, Malaysia, pp. 129–36.

Subhadrabandhu, S. and Kaiviparkbunyay, K. 1998. Effect of paclobutrazol on flowering, fruit setting and fruit quality of durian (Durio zibethinus Murr.) cv. Chanee, *Kasetsart Journal* (Natural Science Supplement) 32: 73–83.

Subhadrabandhu, S., Schneemann, J.M.P. and Verheij, E.W.M. 1991, *Durio zibethinus* Murray. In E.W.M. Verheij and R.E. Coronel (eds), *Plant Resources of South-East Asia (PROSEA) 2: Edible Fruits and Nuts*, Pudoc, Wageningen, The Netherlands, pp. 157–61.

Subhadrabandhu, S. and Shoda, M. 1977, Effects of time and degree of flower thinning on fruit set, fruit growth, fruit characters and yield of durian (*Durio zibethinus* Murr.) (cv.) Mon Thong, *Kasetsart Journal* (Natural Science Supplement) 31: 218–23.

Subhadrabandhu, S. and Tongumpai, P. 1990, Off-season production of some economic fruit in Thailand. In *Off-season Production of Horticultural Crops*, Book Series No. 41, Food and Fertilizer Technology Centre for the Asian and Pacific Region, Taipei, Taiwan, pp. 78–88.

Sutthaphan, S. 1992, Effects of coating materials 'Semperfresh' and 'Sta-fresh 7055' and low temperature on storability of durian, Undergraduate Special Problem, Department of Horticulture, Kasetsart University, Bangkok, Thailand (in Thai).

Sutthaphan, S. 1993, Post-harvest changes in chemical composition of Chanee and Mon Thong durian pulp, M.Sc. thesis, Kasetsart University, Bangkok, Thailand (in Thai with English abstract).

Tai, L.H. 1971, Studies on *Phytophthora palmivora*, the causal organism of patch canker disease of durian, *Malaysian Agricultural Journal* 48: 1–9.

Techniques of orchard management for durian and rambutan 1989, *House Agricultural Magazine*, Special Issue, Charoenrut Publishers Ltd, Bangkok, Thailand (in Thai).

Terra, G.J.A. 1952, Some ecological requirements of Indonesian fruit trees, *Landbouwen* 24: 193.

Thailand Institute of Science and Technology 1997, The production and the utilisation of durian powder for health foods, Bangkok, Thailand (in Thai).

Thompson, A. 1934, A disease of the durian tree, *Malaysian Agricultural Journal* 22 (8): 369–71.

Thompson, A. 1938, A root disease of the durian tree caused by *Pythium complectens* Braun., *Malaysian Agricultural Journal* 26 (1): 460–4.

Tinggal, S. 1992, *Brunei Darussalam Fruit in Colour*, Star Trading and Printing Co. Ltd, Bandar Seri Begawan, Brunei Darussalam.

Tongdee, S.C., Chayasombat, A. and Neamprem, S. 1988a, Respiration, ethylene production and changes in the internal atmosphere of durian (*Durio zibethinus* Murray). In Proceedings of the Seminar on Durian, 25–26 February 1988, Thailand Institute of Scientific and Technological Research, Bangkok, Thailand, pp. 22–30 (in Thai).

Tongdee, S.C., Chayasombat, A. and Neamprem, S. 1988b, Effects of harvest maturity on respiration, ethylene production and the composition of internal atmospheres of durian (*Durio zibethinus* Murr.). In Proceedings of the Seminar on Durian, 25–26 February 1988, Thailand Institute of Scientific and Technological Research, Bangkok, Thailand, pp. 31–6.

Tongdee, S.C. and Suwanagul, A. 1988. Effect of low oxygen atmospheres on ripening in durian (Durio zibethinus Murray). In *Proceedings of the Seminar on Durian*, pp. 25–26 February 1988, Thailand Institute of Scientific and Technological Research, Bangkok, pp. 37–45 (in Thai).

Tongdee, S.C., Suwanagul, A., Neamprem, S. *et al.* 1990, Effect of surface coatings on weight loss and internal atmosphere of durian (*Durio zibethinus* Murr.) fruit, *ASEAN Food Journal* 5: 103–7.

Valmayor, R.V., Coronel, R.E. and Ramirez, D.A. 1965, Studies on floral biology, fruit set and fruit development in durian (*Durio zibethinus* Murr.), *Philippines Agriculture* 58: 355–66.

Vangnai, V. 1983, Species and varieties of Thai fruits, Department of Horticulture, Faculty of Agriculture, Kasetsart University, Bangkok, Thailand (in Thai).

Vangnai, V. 1996, Status report on genetic resources of durian in Thailand, Project No. B06, International Plant Genetic Resources Institute, Singapore.

Veijjajiva, P. 1986, Durian for export, *House Agricultural Magazine* 10 (110): 9–15 (in Thai).

Vietmeyer, N.D. 1975, Underexploited tropical plants with promising economic value, National Academy of Science, Washington D.C., USA.

Watson, B.J. 1978, *An Appreciation of Five Exotic Tropical Fruit Tree Species and their Potential in Queensland*, Kamerunga Horticultural Research Station, Department of Primary Industries, Brisbane, Queensland, Australia.

Watson, B.J. 1984, Durian (*Durio zibethinus* Murr.). In P.E. Page (ed.), *Tropical Tree Fruits for Australia*, Department of Primary Industries, Brisbane, Queensland, Australia, pp. 45–50.

Wester, P.J. 1912, New Philippine fruits, *Philippines Agricultural Review* 5: 592–7.

Wilson, T.B. 1954, The marketing of durian fruit from Parit District, Perak, *Malaysian Agricultural Journal* 37 (4): 211–17.

Yaacob, O. 1983, The growth pattern and nutrient uptake of durian (*Durio zibethinus* Murr.) on an oxisol, *Soil Science and Plant Analysis* 14 (8): 689–98.
Yaacob, O. and Subhadrabandhu, S. 1995, *Production of Economic Fruits in South-East Asia*, Oxford University Press, Kuala Lumpur, Malaysia.
Yu, Y.B., Adams, D.O. and Yang, S.F. 1980, Inhibition of ethylene production by 2,4-dinitrophenol and high temperature, *Plant Physiology* 66: 286–90.
Zainal Abidin, M. 1992, New hybrid durian clone. In Proceedings of the Seminar on Durian, 25 June 1992, Malaysian Agricultural Research and Development Institute, Ipoh, Perak, Malaysia.
Zainal Abidin, M. 1993, New durian hybrids MDUR 78, MDUR 79 and MDUR 88, Teknological Buah-buahan, No. 9, Ministry of Agriculture, Kuala Lumpur, Malaysia (in Malay).
Zainal Abidin, M., Idris, S. and Aman, M. R. 1996, Status report on genetic resources of durian in Malaysia, Project No. B06, International Plant Genetic Resources Institute, Singapore.

Index

D

Q

R

S

Printed and bound by CPI Group (UK) Ltd, Croydon, CR0 4YY
16/09/2026
14963437-0001